A Research-Oriented Laboratory Manual for First-Year Physics

A Manual that Incorporates a Semester-Long Research Project into the First-Year Physics Curriculum

Chris McMullen, Ph.D.

A Research-Oriented Laboratory Manual for First-Year Physics

www.lsmsa.edu/faculty/CMcMullen

Custom Books

Textbooks / Science / Physics

ISBN: 1440404143

EAN: 9781440404146

Contents

Index to Handy Notes

Research Timeline

	Part 1		Part 2
Expt.	Research Component	Expt.	Research Component
1	Choose topic, find a reference	15	Choose topic, find a reference
2	Prepare introduction	16	Prepare introduction
3	Write formalism	17	Write formalism
4	Draw illustrations	18	Draw illustrations
5	Develop algorithm, add references	19	Develop algorithm, add references
6	Implement algorithm, begin discussion	20	Implement algorithm, begin discussion
7	Continue algorithm, make graphs	21	Continue algorithm, make graphs
8	Complete algorithm, write abstract, complete discussion	22	Complete algorithm, write abstract, complete discussion
9	Check algorithm, draw conclusions	23	Check algorithm, draw conclusions
10	Submit paper, review a paper	24	Submit paper, review a paper
11	Peer review process, submit revised paper	25	Peer review process, submit revised paper

To the Instructor

Overview: This laboratory manual integrates a strong research component into the first-year physics laboratory curriculum. The motivation for this originates from a growing trend among physics departments to involve undergraduate students in research projects earlier in their academic pursuits. The primary goals of this research-oriented laboratory curriculum are to prepare first-year students for prospective undergraduate research, inspire student interest in undertaking a research project, and provide students the opportunity to gain research experience during their first-year physics course.

Research Project: One feature of this research-oriented curriculum is that it allows students to work on a research project in a way that is not overbearing to a student seeking a lone lab credit. Similarly, the research projects need not entail many extra hours of mentoring and grading from the effort of the instructor or teaching assistant.

This manual is set up such that students will submit various components of their research projects throughout the semester so that by the end of the semester the project is complete, thereby removing the burden of an overwhelming assignment due at the end of the semester. The usual lab report time can be trimmed each week to make time for research.

The process of identifying the purpose, describing the procedures, tabulating data, performing calculations, analyzing results, determining errors, contemplating sources of error, discussing results, and drawing conclusions for each lab can quickly become tedious. A few diligent students will wear themselves out, while many will never put forth a respectable effort. Recognizing this, it is common to require a few formal reports, in which a better effort is expected; and if students still turn in informal reports for other experiments, the problems associated with the tedium of lab reports has not been remedied. Still, students are learning to write all of sections of the report simultaneously, and similarly instructors or their teaching assistants receive a thick pile of reports at once. The challenge in this platform is to help each student become more adept at writing each component of the report.

A different prescription is suggested in this research-oriented curriculum: Collect just one or two *written* sections each week. Students will focus on how to write one or two sections at a time. They should be encouraged to put more effort into it – it helps to warn that if the class lacks effort, it is not too late to revert to the usual method of collecting full reports each week – and hopefully learn more from the experience.

These focused lab reports set aside time that can be put into a research project. Students turn in one component of their research project each week – in addition to their lab report – culminating in a formal research paper at the end of the semester. Instructors with relatively long lab periods – say, 2.5 to 3 hours – should consider collecting the lab reports at the end of lab, so that students do not have to work on both lab reports and research prior to the next lab meeting.

Compared to the usual practice of collecting weekly lab reports, this research-oriented laboratory curriculum offers students the opportunity to receive feedback and make revisions for one ongoing project, instead of always moving onto the next lab.

Depending upon the specific needs of the instructor, the research projects may be assigned as optional rather than mandatory of every student in the class. The case of the optional project will not solicit much participation, however, if the research project is simply perceived as extra work. One way to motivate optional research from interested students is to advertise that the work load will be moderately balanced between those on the traditional and research options. This is possible in this platform, whereby students on the traditional option can simply turn in extra sections of a lab report in lieu of the research component that is due each week; in this way, they develop a formal report at the end of the semester instead of a research paper. Another possibility is to offer two different syllabi with an incentive to those who choose to do research – e.g. a lab final for the traditional option to substitute for the final research paper.

A list of several possible research topics suitable for a first-year physics student is provided for convenience. This list is restricted to computer-based projects, since they do not involve the problem of students needing to arrange time to use equipment in lab in order to conduct their research and can be conducted by students without the need of supervision. Since most real-world research is computational, a computer-based project is a worthwhile endeavor. The projects do not require mastery of sophisticated techniques, and many can be completed using a spreadsheet such as Microsoft Excel, for the benefit of those students who lack programming skills. The students are not seeking novel results to publish in a journal, but a project that relates to concepts from the course to serve as the basis for introductory research experience. The scope and details of the various projects can be modified to suit the needs of an individual instructor, or the project ideas listed may serve to inspire other ideas.

Some students may not satisfy all of the theoretical and numerical challenges of their projects. The goals set marks for the students to strive toward, which some students will reach. The research components that the students submit on a weekly basis will be helpful for assessing individual efforts and progress, which will aid in the assignment of grades for unfinished projects.

Research Preparation: The research project is not the only way that this manual is oriented toward research. The sections that are collected each week closely parallel a journal article – e.g. abstract instead of purpose statement, or an experimental section describing how the equipment was used different from how a set of procedures outlines what to do. The first time a section is required, it is described in detail to help students learn how to write it. The final paper turned in at the end of the semester – which could be a formal lab report, if not a research paper – takes the shape of a journal article. A collaborative peer review session at the end of the semester, in which students share their papers and input, is a process that mimics the peer review process and prepares them to handle and offer constructive criticism.

Prelab exercises are written with potential research skills in mind, such as an assignment that requires looking information up in the library and citing the reference, or entails reading theory and applying it to the experiment, or designing an experiment to meet a given set of objectives or to test a hypothesis (so students will practice developing a set of procedures before the experiment, not just summarizing them afterward). This latter example is a part of inquiry-based learning, which is utilized here to some degree. Here, value is placed on the notion that students need to see some structure and examples of procedures early on, and then promote more independent experimentation and design as the course progresses.

Experiments: The first eleven experiments of Parts 1 and 2 include instructions for what should be done or submitted in order for students to complete their research projects by the semester's end, since many lab courses perform a dozen experiments in a semester. A few extra labs have been included for those with extra weeks; others may opt to skip one of the earlier labs in preference for one of the last labs.

Almost all of the experiments involve standard equipment available in most physics labs. If desired, one or more experiments may be substituted. For the most part, the ordering of the labs can be changed to suit the needs of the instructor without significant pedagogical harm. In the case of substituting or skipping an experiment, students will still need to work on the component of their research specified in the lab write-up for the missed experiment.

Lab Write-Ups: Standard theory, such as how to add vectors or conserve energy, is not provided in the lab write-ups for a couple of reasons. For one, there is already a thorough introduction to the standard theory in the textbook, which is also covered during the lecture and applied to the problems. Each write-up does have a section that outlines which concepts should be read in preparation for lab and which equations will be useful. Secondly, a useful part of research is searching for what is already known on the topic, then reading, understanding, and applying it. Students can practice this through the referrals to relevant sections of the textbook; but not specific sections – students may need to learn how to use the index and table of contents. Invariably, the lab will run ahead of the lecture, largely since it would otherwise be impossible to learn material at the end of the course and then perform a related experiment. When this occurs, students will be reading material in the textbook – reading the textbook, that's a good thing! – that has not already been covered in class, and may need to apply it to complete prelab exercises. Some students will be frustrated by the situation, unless they are led to see the potentially useful experience gained from this. Besides, they are not being left completely in the dark: The beginning of the lab period is generally dedicated toward reviewing the theory. In this way, the students practice learning something independently, and then they can relearn it properly through the instructor's aid.

Each lab does have a short introduction. The introduction carefully leaves out equations that are intended for students to gather during their related textbook readings. However, in the rare case that the material and equations are deemed not to be so standard, they are presented in the introductions.

The objectives do not list a myriad of skills to be learned from each lab, but focus on one or two, usually, main experimental goals. The experiments are oriented around these objectives, leading up to some labs where students have the opportunity to try their own hands at designing and conducting experiments to meet a specified set of objectives.

In a few cases, the apparatus lists alternative equipment, for the benefit of instructors who may not have the same equipment available. The procedures do not provide cookbook instructions, but aim to guide the development of how to design and conduct an experiment. A useful Data Check summary highlights which quantities must be recorded prior to leaving lab.

The Analysis section describes the calculations, graphs, and quantitative error analysis. This is where the report begins. Students are asked to organize the key results into a concise table, forcing them to sort through their calculations for what is important.

Each write-up concludes with a visible section on what to turn in, starting with which sections should be included in the report. This guide focuses on one or two written sections per lab, as explained earlier, but is a suggestion that can be modified to suit the specific needs of each instructor. The research section helps to guide students along in their projects, making the project manageable by completing it little by little throughout the semester.

Finally, prelab exercises, described earlier, appear at the end of the lab write-up prior to the relevant experiment. Students should complete these prior to lab, but should also read the coming lab write-up and relevant sections of the textbook in preparation for the experiment.

To the Student

Overview: Physics is taught through the combination of the lecture and laboratory. After all, physics is an experimental science. The lecture course aims to develop a firm grasp of the fundamental concepts and problem-solving strategies which form the basis of theoretical physics. Most of the grade, or at least course credit, stems from the lecture owing to the challenging nature of the concepts, which are counterintuitive to most students, and the mathematics involved in solving the problems. However, the laboratory experience more closely resembles real-world physics, captures the spirit of scientific inquiry, and develops a greater variety of skills, which are likely to be very useful to students who do not become physicists – such as structuring arguments in a way that statements are backed up by evidence, making careful measurements, identifying primary sources of error, working with spreadsheets, presentation skills, collaborating with others, and applying error analysis to quantify how well a result agrees with a theoretical prediction. Put more into the laboratory experience... and get more out of it.

This laboratory manual is oriented toward developing the skills necessary to conduct independent, scientific research. This is not a cookbook course where explicit instructions are provided to train students to learn how to follow instructions. Rather, it is a problem-solving course where students are required to think about the concepts, make connections between theory and experiment, solve experimental problems, and develop independence by writing procedures and conducting experiments in order to meet a given set of objectives. Well, that is the goal; more specifics are provided where it is deemed instructive, while more independence is expected as the course progresses.

Responsibility: Conducting experiments requires using apparatus, of which some are potentially dangerous or easily damaged if not utilized appropriately. As such, it is important to behave responsibly in the laboratory. This entails:

- coming to each experiment on-time and prepared; otherwise, accidents or damage to equipment may result from lacking key information
- inspecting the equipment before using it and first consulting the instructor if it is damaged or missing articles
- following instructions and heeding warnings during lab
- using equipment responsibly
- keeping the lab tidy in an effort to prevent accidents
- reporting any accidents to the instructor immediately
- cleaning the lab station and putting equipment away neatly before leaving

Complete the prelab exercises in advance and bring them ready to submit when lab begins. Also, read the lab write-up ahead of time, in addition to relevant sections from the textbook and class notes in order to conduct the experiment more efficiently and enhance understanding of the lab. Lack of preparation for lab or alertness during the experiment is not only a safety hazard, but is unfair to lab partners.

Part 1: Mechanics

First-Semester Sample Research Topics

Described below are 25 research projects that are extensions of topics from first-semester physics. The projects vary somewhat in challenge and complexity. Each problem includes one or more numerical goals that can be achieved through computer programming; many of these can also be solved using a spreadsheet program such as Excel. You must write the code yourself – you may *not* use part of a program written by someone else.

1. **Motion Graphs**. The techniques of analyzing motion through graphs of position, velocity, and acceleration apply very generally, whereas the equations of uniform acceleration are limited to a much smaller class of problems in which the acceleration is constant. Theoretical goals: Prepare a theory section that outlines the prescription for how to analyze motion graphs, which connects the equations of motion to the techniques. This should include finding net displacement, total distance traveled, average speed, average velocity, instantaneous velocity, average acceleration, and instantaneous acceleration from graphs of position, velocity, or acceleration. Numerical goals: Given the equation for a curve, write a numerical algorithm that can find the slope of the tangent line at any point or find the area under the curve over a specified interval by dividing up the time axis into a given number of tiny chunks. Apply this toward the numerical analysis of a graph of position, velocity, or acceleration as a function of time.

2. **Vector Addition**. Average wind direction can be determined via vector addition. Theoretical goals: Write a prescription for finding the magnitude and direction of the resultant of a large number of given vectors. Look up the equation for the average power harnessed by a windmill in terms of both the wind speed and direction. Apply your prescription for adding vectors to determine the average power harnessed by a windmill over a large number of days. Numerical goals: Write a numerical algorithm to determine the average wind direction in a windy city using one month of data (this can be looked up) for wind speed and direction. Placing a windmill in the direction of the resultant serves as a useful prediction for how to best harness the wind's energy in that city. However, a slightly different answer may improve the windmill's output a little. To see this, start a new problem: Using the same data, assume that the windmill is pointed east and compute the total wind power for the month. Now vary the direction a little and repeat the calculation. Vary the angle in small increments over an entire circle and compare the results to see which angle yields the greatest total wind power harnessed. Compare to the vector addition result.

3. **Maximum Range from a Moving Source**. The angle at which a projectile should be launched in order to maximize the range depends upon the relative motion of the source and landing area. Neglect air resistance, the earth's rotation, and the variation of gravity with altitude. Theoretical goals: Work through the projectile motion strategy symbolically to find the range of a projectile launched from a moving ship in terms of the initial speed of the projectile relative to the ship, the speed of the ship relative to the water, the launch angle, gravitational acceleration, and the initial height of the projectile above the water. Numerical goals: Write a numerical program to compute the horizontal range of a projectile relative to a ship in terms of the same parameters considered in the theory. Run the program for angles varying from $-90°$ to $90°$ with small increments to determine which angle maximizes the range. Make a graph of the angle that maximizes the range as

a function of the relative velocity between the ship and the water, verifying the result when the relative velocity equals zero.

4. **One-Dimensional non-Uniform Acceleration**. The problem of non-uniform acceleration can be solved quite generally with numerical techniques. Theoretical goals: Illustrate the numerical technique by working out an example for a small number of intervals. Compare the results for one interval, two intervals, four intervals, and eight intervals. Check that these values agree with the results of the numerical analysis. Numerical goals: Given a time-dependent equation for acceleration with known values for the initial velocity and initial position, write a program to determine the position, velocity, and acceleration at a specified time using the following prescription. First, compute the initial acceleration. Use the initial acceleration to compute the position and velocity after a short time interval under the approximation that the acceleration is uniform. Now compute the acceleration after that short time interval. Use this acceleration to compute the position and velocity after a second short time interval, plugging in the previous position and velocity as initial values. Continue to shift the acceleration, position, and velocity over repeated time intervals. Graph the results for illustrative examples.

5. **Static Equilibrium**. The conditions for equilibrium can be applied to determine the equilibrant – i.e. the balancing force – for a given situation. Theoretical goals: Prepare a prescription for calculating the equilibrant for an arbitrary number of given magnitudes and directions of forces. Work out an example of the numerical program, showing that the calculation agrees with the numerical analysis. Numerical goals: Write a program that asks the user how many forces there are, asks for the magnitude and direction of each force as input, and uses this to compute the magnitude and direction of the equilibrant. Adapt this program to various applications, such as a boy hanging from a clothesline – given the angles, the program can compute the tensions in each segment of the clothesline, with a little modification.

6. **Air Resistance**. The force of air resistance is a resistive force that is proportional to the speed. In many situations, it is approximately equal to $F = -bv$. Theoretical goals: Apply Newton's second law to solve for acceleration. Illustrate the numerical technique by working out an example for a small number of velocity intervals – namely, dividing the difference between the initial and final velocity in one, two, four, and eight intervals, and comparing. Check that these values agree with the results of the numerical analysis. Numerical goals: Adapt the numerical technique described in the One-Dimensional non-Uniform Acceleration project to the case of air resistance. The acceleration will equal a constant (gravitational acceleration) minus a contribution from the force of air resistance. Since the acceleration is expressed explicitly in terms of the instantaneous speed, rather than the time, work with short velocity intervals instead of short time intervals.

7. **Variation of Gravity with Altitude**. This is the same as the Air Resistance project, except that variation of gravitational acceleration with altitude is accounted for instead of air resistance. In this case, application of Newton's second law will lead to an acceleration that explicitly depends upon position. Work with position intervals, rather than time or velocity intervals.

8. **Uniform Circular Motion**. The equation for centripetal acceleration for an object undergoing uniform circular motion, as well as the direction of the acceleration in this case, can be derived from the definition of instantaneous acceleration in terms of a limit as the time interval approaches zero.

Theoretical goals: Look up the described derivation in a textbook, study it, and present a similar derivation in your own words with your own choice of symbols and labeled diagrams. More detailed explanations are expected than are given in the textbook. Cite your source, as always. Numerical goals: Write a program for an object traveling with uniform circular motion that does the following. The user inputs the radius of the circle, the speed, and two angles to consider; each angle uniquely specifies a point on the circle. Treat the two angles as initial and final positions. Compute the magnitude and direction of the initial and final velocities. Perform vector subtraction of these initial and final velocities, obtaining the magnitude and direction of the change in velocity. Divide the magnitude of the change in velocity by the time interval to compute the magnitude of the average acceleration. The direction of the average acceleration equals the direction of the change in velocity. Show that the smaller the angle between the specified points, the closer the magnitude of the average acceleration agrees with the equation for centripetal acceleration and the closer the direction of the average acceleration is to being centripetal.

9. **Work Done by a Varying Force**. The equation $W = Fs\cos\theta$ is a special case – it applies when the force and angle are constant. However, some forces are not constant. For example, an object's gravitational attraction to a planet depends on its distance from the center of mass of the planet. Theoretical goals: Find a textbook that describes the work done by a varying force. Prepare a theory section that introduces this topic, and explains how to compute the work in this case by comparing with the techniques for analyzing motion graphs. Identify specific analogies with the motion graphs. Work through an example of a graph of a variable force, computing the work done. Check that this agrees with the numerical program. Numerical goals: Write a program that applies the relevant numerical techniques from the Motion Graphs project to compute the work done by a varying force. Input should include the equation for the variable force as a function of position.

10. **Nonconservative Work**. Work done by friction is nonconservative. Nonconservative work is path-dependent. Theoretical goals: Write out conservation of energy symbolically for a block sliding down an incline from rest, where there is friction between the block and the incline. Work through a numerical example, computing also the effect that friction has on the outcome. Set up conservation of energy for a curved, rather than straight, path – e.g. a slide in the shape of a quarter-circle. Explain, precisely, why this problem is much more difficult to solve algebraically. Numerical goals: Given the equation for a curve, the endpoints of the motion (starting from rest), and the coefficient of friction between the block and the curved surface, write a numerical program to determine its final speed. The program should use conservation of energy, accounting for the energy lost in the form of nonconservative work done by the friction force. A nontrivial aspect of this problem is that the normal force depends upon the speed as the inward components of the forces contribute toward centripetal acceleration. Assume that a curve is specified in which the block does not lose contact with the curved surface. Depending upon the initial slope and the coefficient of friction, the block may not slide at all – it may be worth checking this or preventing this, else the program may encounter problems trying to squareroot negative numbers. The friction force depends on the normal force, which depends on the angle that the tangent line makes with the horizontal. This angle can be approximated as follows: Use a small increment Δx at the block's present position, use the equation of the curve to determine the corresponding Δy, and use these to compute the slope of the tangent line – the smaller Δx, the better the approximation, but not past the numerical limitations of the program. This angle that the tangent line makes with the horizontal will vary with the block's position. The nonconservative work depends on the total

distance traveled, or arc length. This can be computed by finding the hypotenuse of the triangles of Δx and Δy used to find the instantaneous angles, and tallying these up to approximate the arc length. Check that your program agrees with the numerical example in the theory section for the case of a straight incline.

11. **Two-Dimensional Collisions**. When a moving billiard ball collides with a stationary billiard ball off-center, the result is a two-dimensional collision. Neglect English and assume the collisions to be elastic. Theoretical goals: Write down the appropriate conservation laws for a two-dimensional elastic collision between two billiard balls, where one billiard ball is at rest prior to the collision and the billiard ball that was at rest travels along the line connecting the centers of the two balls at impact. Strive to solve for the magnitudes and directions of the final velocities in terms of the magnitudes and directions of the initial velocities and the distance for which the collision is off-center. Carry the calculation as far as you can. Numerical goals: Given the radius of a billiard ball, the speed of a moving billiard ball, and the distance for which the collision is off-center, write a program to determine if there is a collision (not worrying about possible rebounds), and, if so, compute the magnitudes and directions of the final velocities. Apply this program to make some examples of where a cue ball should be aimed in order to knock another ball into a particular pocket.

12. **Center of Mass of a System of Particles**. Given the coordinates for a system of pointlike objects lying in a plane, there is a mathematical prescription for determining the center of mass of the system. Theoretical goals: Prepare a detailed description for how to determine the center of mass of a system of numerous pointlike objects, given their coordinates and masses. Illustrate this with some numerical examples. Numerical goals: Write a program where a user is asked how many objects are in the system, accepts the mass and coordinates as input for each object, and computes the center of mass of the system. Check that this agrees with the examples from the theory section. Additionally, make a variation of the numerical program with the following adaptation: Apply the algorithm to compute the center of mass of a system consisting of a boy with a given mass at a certain position on a canoe of a given length and mass. Conserve momentum as the boy walks across the canoe with a specified speed, and see what happens to the center of mass as the boy walks across the canoe.

13. **Center of Mass of a Rigid Body**. The center of mass of an object with a continuous distribution of mass, rather than a discrete system of pointlike masses, can be found following the prescription outlined in the Center of Mass of a System of Particles project by dividing the object up into a large number of tiny rectangles and treating each rectangle as a pointlike object (it is a straightforward matter to determine the mass of each rectangular chunk). Theoretical goals: Take an object shaped like an L or an F, for example, and show how to find the center of mass of the object by dividing it into rectangular chunks. Explain how the same process can be applied to find the center of mass of a solid disc, for example, and discuss approximations and errors. Numerical goals: Write a program that computes the center of mass of a solid 2D object from the equation of a closed curve (such as an ellipse, or semicircle + line) by dividing the mass up into a large number of tiny chunks.

14. **Center of Gravity**. This is similar to the Center of Mass of a Rigid Body project, except for computing center of gravity instead of center of mass. Apply your numerical algorithm to

determine how far the center of gravity of the Empire States building is from its center of mass, accounting for the fact that the Empire States building is 3D. (Don't worry about the canoe part.)

15. **Torque**. This is similar to the Static Equilibrium project, except for finding the balancing torque instead of the balancing force. It is necessary to input the position where the force is applied, in addition to the number, magnitude, and direction of the input forces. Additionally, make a variation of the numerical program with the following adaptation: Compute the net torque exerted on a meterstick, given the input forces, for the case where the fulcrum is off-center. It will be necessary to specify the location of the fulcrum as well as the mass of the meterstick. (Don't worry about the clothesline part.)

16. **Maximum Load**. The components of the net force and the net torque are all zero for a rigid body in static equilibrium. Theoretical goals: Write a theory section that describes the problem-solving strategy for static equilibrium. Work through some numerical examples that illustrate the technique. Numerical goals: Write a numerical program that solves static equilibrium problems involving a boom, load, hingepin, and tie rope. Input will require the endpoints of the boom (allowing for tilt), the exact position on the boom where the load is suspended, which endpoint of the boom is connected to a vertical wall by a hingepin, the point on the boom where the tie rope is connected, the point on the vertical wall to which the other end of the tie rope is connected, the mass of the boom, and the mass of the load. The program needs to compute the tension in the tie rope and the magnitude of the force exerted on the hingepin.

17. **Moment of Inertia for a System of Particles**. This is similar to the Center of Mass of a System of Particles project, except for computing moment of inertia instead of center of mass. It will also be necessary to specify the axis of rotation. Also, use the program to verify the parallel-axis theorem. (Don't worry about the canoe part.)

18. **Moment of Inertia for a Rigid Body**. This is similar to the Center of Mass of a Rigid Body project, except for computing moment of inertia instead of center of mass. It will also be necessary to specify the axis of rotation. Also, use the program to verify the parallel-axis theorem. Investigate only 2D objects, like ellipses and semicircles.

19. **Rolling without Slipping**. Look up the formulas for the moment of inertia of a rotating sphere and cylinder about the natural rolling axes for the case where the object has a finite thickness and hollow cavity in the center. This is different from the more common formulas where the objects are infinitesimally thin. Theoretical goals: Symbolically, derive equations for the tangential acceleration and final tangential speed of a rigid body rolling along an inclined plane, given whether it is a sphere or a cylinder, the ratio of its inner radius to its outer radius, its initial tangential speed, and the distance traveled down the incline. Numerical goals: Write a numerical program that solves problems with such hollow spheres or cylinders rolling without slipping down an inclined plane. Figure out how many input parameters must be specified (and which combinations provide sufficient information) and let the user choose what to input. The program will compute the remaining unknowns. Apply this to numerical examples to check your work.

20. **Conservation of Angular Momentum**. The angular momentum of a system is conserved if the net external torque equals zero. Theoretical goals: Write a theory section that introduces the

concepts and problem-solving strategies associated with the conservation of angular momentum, beginning with what angular momentum is. Draw on some analogies with linear motion. Work through a couple of instructive numerical examples. Numerical goals: Write a program that conserves angular momentum for a system of rotating rigid bodies to compute the final angular speed of the system. The input should allow for an arbitrary number of rigid bodies, accommodate a variety of shapes of rigid bodies, accommodate a variety of examples (including collisions with pointlike masses), and require the mass, size, shape, position, and initial angular velocity (some of which may be zero) of the rigid bodies (and any pointlike objects). Check that the program agrees with the examples from the theory section.

21. **Kepler's Laws**. Kepler's laws describe the orbits of satellites. Theoretical goals: Apply the satellite strategy to derive Kepler's third law for the special case of a circular orbit. Set up conservation of energy for a satellite in an elliptical orbit, noting that the potential energy is not mgh since gravitational acceleration is not constant in this case (the correct formula can be looked up). Describe the energy transformations that occur during one period. Set up conservation of angular momentum, and explain how it relates to Kepler's second law. Numerical goals: Write a program to plot a planet as it completes an elliptical orbit, accounting for the fact that the speed of the planet varies with its distance from the sun (at one focus). This requires the ability to write a program that can make an animated plot. Use Kepler's second law and the equation for the ellipse to correctly determine how to vary the speed as the planet's position varies. Now add planets to your graph. The computer program needs to determine the relative speeds of the planets needed in order to satisfy Kepler's third law, while also varying the speed of each planet appropriately to satisfy Kepler's second law. User input should include the relative distances of the planets.

22. **Oscillating Spring with Friction**. Consider a horizontal spring connected to a wall at one end and a block at the other. When the system is experiencing oscillation, friction between the block and the horizontal surface will eventually bring the system to rest. Theoretical goals: Apply Newton's second law to derive a symbolic equation for the acceleration of the block. Conserve energy to symbolically relate the speed of the block to the position of the block. Describe changes in the speed and acceleration of the block as the block oscillates, referring to your derivations to support your statements. Explain how these equations, compared to the case of a frictionless surface, affect the motion of the system. Numerical goals: Apply the numerical technique described in the One-Dimensional non-Uniform Acceleration project to determine how the speed and acceleration vary with position, and also predict when the block will change direction and account for this in the program. Vary position, rather than time, in short intervals since the equation for acceleration depends on position, rather than time, explicitly. Plot speed and acceleration as functions of position; these may be multi-valued functions since the block is oscillating.

23. **Simple Pendulum**. Compared to the period of a spring, the typical textbook equation for the period of a simple pendulum involves a simplifying assumption concerning the angle of oscillation. Theoretical goals: Look up the derivation of the period of a simple pendulum. Study this, and prepare a derivation in your own words, with your own choice of symbols and labeled diagrams. Work through a numerical example that illustrates your theoretical results. Numerical goals: Start with the equation from the theory that involves angular acceleration and angle, but do not make the aforementioned simplifying assumption. Apply the numerical technique described in the One-Dimensional non-Uniform Acceleration project to determine how the angular position, angular

speed, and angular acceleration depend on time. It may be convenient to vary angle, rather than time, in short intervals since the angular acceleration explicitly depends on angle in the starting equation, but if so, you will still need to involve time in the calculations to determine the time-dependence of the quantities. Notice that you are working with the angular analogs to the linear motion variables.

24. **Buoyancy**. Archimedes' principle describes the buoyant force that an object experiences when it is partially or wholly submerged in a fluid. This project is similar to the Static Equilibrium project, except that an object must be at least partially submerged in a fluid (like an ice cube floating at the top of a cup of water).

25. **Heat Engines**. The thermodynamic processes involved in a heat engine can be displayed in a *PV* diagram, which can be analyzed with graphical techniques. Theoretical goals: Read about various heat engines. Study these heat engines, then prepare an introduction to the theory of heat engines and how *PV* diagrams are related to the concepts involved. Explain how to interpret a *PV* diagram, and illustrate this by working through a numerical example. Numerical goals: Given equations for the curves in a *PV* diagram, first compute the points of intersection to determine the endpoints of each path. Then compute the work done for each cycle of the diagram. Compare the results to the numerical example worked in the theory. The area under a curve can be numerically approximated by dividing the region into thin rectangles.

Experiment 1: Measurement and Uncertainty

Introduction: Aside from counting a reasonable number of objects, the measurement of any physical quantity – such as length, force, or time – involves inherent experimental errors. As an example of this, consider a marble dropped from the roof of a building. The time of its descent, if measured with a stopwatch, will not be *exact*. An obvious reason is that only a few significant figures are displayed on the stopwatch, so the infinite number of trailing digits would have to *all* be zero for the measurement to be exact. However, this is an insignificant source of error, since it pales in comparison with a greater relative error – the reaction time of the person making the measurement. A very important analysis skill is the ability to identify sources of error and correctly determine which are most significant.

This measurement could be improved by utilizing better equipment. For example, electronic sensors may detect when the marble is released and when it strikes the ground. Yet, no matter how good the sensors are – or how good they could be with an unlimited budget and time to develop better technology – the measurement will never be exact, and not because of limited precision of the measuring device. Consider some of the problems that occur at the atomic scale. Electrons are in motion around nuclei, there is not clear contact in the interactions between the elementary particles involved to define the moment of impact exactly, and at this scale Heisenberg's uncertainty principle becomes important.

It is necessary to accept that physical measurements involve some error, and to realize the importance of quantifying how good a measurement is. Although it is not possible to measure a quantity exactly, it is possible to assess how close it is to the actual value. To see the importance of this, imagine that an acquaintance wants to sell you a chunk of metal for $100 that looks and feels like gold. If it turns out to be gold, your investment will make you rich (it's a large chunk), but if it turns out to be worthless, you'll be out a hundred dollars. To aid your decision, you measure the density, which you determine to be 18.5 g/cm^3. Then you look up the accepted value for the density of gold, which is 19.3 g/cm^3. Should you invest?

Whether or not you would be making a sound investment depends upon the uncertainty in your calculated density. If the uncertainty is ± 2.0 g/cm^3, then the measurements do agree within the uncertainty, since 19.3 is greater than 16.5 and less than 20.5. However, over this range, the chunk could also turn out to be another metal or a combination of metals. Yet, there is a chance, so the gamble *could* pay off. If the uncertainty is instead ± 0.2 g/cm^3, then there is a clear disagreement. In this case, the purchase would be a very poor investment. Taking time to quantify the uncertainty is critical to establishing the level of agreement between two results.

Textbook Reading: Read the sections of the textbook on measurement, systems of units, SI units, scientific notation, significant figures, and, if available, uncertainties and error propagation.

Objective: Given the dimensions of a rectangular sheet of paper, theoretically predict and experimentally determine the area and perimeter after repeatedly folding the sheet widthwise a specified number of times.

Apparatus: ruler, vernier calipers.

Laboratory Notebook

Bring a dedicated, bound laboratory notebook, such as a composition journal, to lab each week. This notebook serves as a place to record qualitative observations and quantitative data, note important points during lab that will be useful for writing the report later or for redoing the experiment at another time, sketch equipment, and take notes during lab.

Procedures: Use a ruler to measure the length of a rectangular sheet of paper. Use the metric system, and record the measurement to the nearest 0.1 mm, which requires estimating between the millimeter markings. Measure the length in a couple of other positions; these lengths may differ slightly from the first length measurement. Now measure the width of the sheet in a few different positions.

Recording Data

A good experimentalist records data directly into a laboratory notebook with non-erasable ink in order to protect the integrity of the experiment. If a mistake is made recording the data, simply line it out with a single line such that the original entry remains legible, and add a brief note to explain the mistake. Record the date on which the measurements are made. All quantitative measurements and qualitative observations should be recorded in the laboratory notebook (not to be confused with this laboratory manual). Also record the first and last names of all partners. Each member of a group must record his or her own data in his or her own notebook. Record the data directly into the notebook when the measurements are made. Do not write it down on a separate sheet of paper to be copied into the notebook later, as this defeats the whole purpose.

Fold the sheet of paper widthwise and repeat the length measurements. Use the vernier calipers for any of the distance measurements that the device can accommodate.

Starting with the already folded sheet, fold it widthwise again, repeating the width measurements. Repeat the widthwise folding and measurement a couple of more times.

Vernier Calipers

Examine the vernier calipers. Like rulers, some vernier calipers read metric on one side and British on the other. Choose metric. A measurement with typical vernier calipers consists of four significant figures – two from the main scale and two more from the vernier scale. Close the vernier calipers completely and note that the zero line – not the left edge – of the vernier scale lines up with the zero line of the main scale when the reading is zero. If it is slightly off, it will be necessary to account for this *zero reading* in all measurements.

To read the vernier calipers, first examine the zero line (*not* the left edge) from the vernier scale to see what marking it is just past on the main scale. This yields the first two digits of a four-digit reading. For example, the zero line may lie in between the 1.1- and 1.2-cm marks. In this case, the first two digits of the reading are 1.1 cm, but this number by itself is unsatisfactory because the measurement is incomplete.

Next, compare the lines of the main scale with the lines of the vernier scale. Most of the lines do not match up, but a few will line up well. Examine the pairs closely to see which pair of lines – one from the main scale and one from the vernier scale – match up best. Read the line from the vernier scale that best matches its partner from the main scale. This yields the last two digits of the four-digit reading. For example, suppose that the second line after the 2 on the vernier scale matches up best with the line above it from the main scale. If there are five subdivisions between the 2 and the 3 on the vernier scale, then these last two digits would be 0.024 (since the second line, in this case, if 40% of the way from the 2 to the 3). The complete reading would be 1.124 cm, putting the two pairs of measurements together.

Data Check: Prior to leaving lab, the following data should be gathered:
- ✓ multiple lengths and widths of the unfolded sheet
- ✓ multiple lengths of the sheet after one fold
- ✓ multiple widths of the sheet after two folds, and repeated data for additional folds

Analysis: Calculate the average length and width of the unfolded sheet as well as the standard deviations in these quantities. Repeat for the folded sheets.

Average and Standard Deviation

When repeated measurements are expected to be approximately equal, the best value for the measurement is the average, defined as

$$\bar{x} = \frac{x_1 + x_2 + \cdots + x_N}{N}$$

while the uncertainty in that quantity is given by the standard deviation, according to

$$\sigma_x = \sqrt{\frac{(x_1 - \bar{x})^2 + (x_2 - \bar{x})^2 + \cdots + (x_N - \bar{x})^2}{N - 1}}$$

where N is the number of measurements made.

Example 1-1: The potential difference, ΔV, across a battery is measured five times: 1.337 V, 1.325 V, 1.296 V, 1.352 V, 1.314. What is the best value for the potential difference and its uncertainty?

The average value of the potential difference is

$$\overline{\Delta V} = \frac{\Delta V_1 + \Delta V_2 + \Delta V_3 + \Delta V_4 + \Delta V_5}{5} = \frac{1.337 \text{ V} + 1.325 \text{ V} + 1.296 \text{ V} + 1.352 \text{ V} + 1.314 \text{ V}}{5}$$

$$\overline{\Delta V} = 1.325 \text{ V}$$

while the standard deviation is

$$\sigma_{\Delta V} = \sqrt{\frac{(\Delta V_1 - \overline{\Delta V})^2 + (\Delta V_2 - \overline{\Delta V})^2 + (\Delta V_3 - \overline{\Delta V})^2 + (\Delta V_4 - \overline{\Delta V})^2 + (\Delta V_5 - \overline{\Delta V})^2}{4}}$$

$$\sigma_{\Delta V} = \{(1.337 \text{ V} - 1.325 \text{ V})^2 + (1.325 \text{ V} - 1.325 \text{ V})^2 + (1.296 \text{ V} - 1.325 \text{ V})^2$$
$$+ (1.352 \text{ V} - 1.325 \text{ V})^2 + (1.314 \text{ V} - 1.325 \text{ V})^2]/4\}^{1/2}$$

$$\sigma_{\Delta V} = 0.021 \text{ V}$$

Units

Always include appropriate SI units after numerical values. Observe that uncertainties have units, too.

Determine the perimeter and area of the unfolded and folded sheets. Propagate errors to derive formulas for the uncertainty in the perimeter and area in terms of the average length, average width, and uncertainties in the length and width. Use the formulas to compute the uncertainties in the perimeter and area. Derive or justify a formula that predicts the area of the rectangular sheet in terms of the original area of the unfolded sheet after folding it N times. Repeat for the perimeter.

Propagation of Errors

When physical quantities are calculated from a formula, the best estimate of the uncertainty is found by a mathematical technique referred to as the propagation of errors. Suppose that N quantities have been measured, $x_1, x_2, \cdots, x_N$, and that their uncertainties, $\sigma_{x_1}, \sigma_{x_2}, \cdots, \sigma_{x_N}$, have already been determined. If a new quantity y is computed in terms of $x_1, x_2, \cdots, x_N$, then the uncertainty in y can be found from error propagation formulas. Two special cases include when the variables are related through addition/subtraction or through multiplication/division.

If y has the form $y = c_1 x_1 \pm c_2 x_2 \pm \cdots \pm c_N x_N$, where $c_1, c_2, \cdots, c_N$ are constant coefficients, then the uncertainty in y equals

$$\sigma_y = \sqrt{c_1^2 \sigma_{x_1}^2 + c_2^2 \sigma_{x_2}^2 + \cdots + c_N^2 \sigma_{x_N}^2}$$

This is referred to as *addition in quadrature*.

If instead y has the form $y = c x_1^{p_1} x_2^{p_2} \cdots x_N^{p_N}$, where c is a constant coefficient and $p_1, p_2, \cdots, p_N$ are constant exponents, then the uncertainty in y equals

$$\sigma_y = y \sqrt{\left(p_1 \frac{\sigma_{x_1}}{x_1}\right)^2 + \left(p_2 \frac{\sigma_{x_2}}{x_2}\right)^2 + \cdots + \left(p_1 \frac{\sigma_{x_N}}{x_N}\right)^2}$$

Example 1-2: A ball rests at a height h above a table, where the height of the table is . Find the uncertainty in the height of the ball relative to the floor.

These derivations are simpler than they may first appear; the various symbols and subscripts can be intimidating at first, but once understood it becomes a useful notation. Essentially, it entails (1) determining whether the formula is more like the addition/subtraction or multiplication/division error propagation case, (2) identifying which symbols are playing similar roles, (3) making a series of substitutions, and (4) simplifying the result, if necessary.

Start with the equation for the height of the ball relative to the floor, which we can call F:

$$F = h + H$$

This is clearly of the addition/subtraction variety. In this case, h and H are measured quantities, corresponding to x_1 and x_2, and F is the calculated quantity, corresponding to y. For this identification to work, the coefficients must equal unity: $c_1 = c_2 = 1$. Substituting these corresponding symbols into the appropriate error propagation formula,

$$\sigma_F = \sqrt{\sigma_h^2 + \sigma_H^2}$$

Example 1-3: The resistivity ρ of a wire with constant cross section is related to the diameter D, length L, and resistance R of the wire through the equation

$$\rho = \frac{\pi D^2 R}{4L}$$

Derive a symbolic formula for the uncertainty in the resistivity.

In this case, D, L, and R are the measured variables, playing the role of $x_1, x_2, \cdots, x_N$, and ρ is the calculated quantity analogous to y. The formula for ρ has the form of the multiplication/division rule, with $p_1 = 2$, $p_2 = -1$, and $p_3 = 1$. The uncertainty in ρ is therefore

$$\sigma_\rho = \rho \sqrt{4\left(\frac{\sigma_D}{D}\right)^2 + \left(\frac{\sigma_R}{R}\right)^2 + \left(\frac{\sigma_L}{L}\right)^2}$$

Results: Make a table of the main quantitative results for this experiment with appropriate units and significant figures. State results in the format $x \pm \sigma_x$. The table of results should be focused on main results, and a *not* be a comprehensive list of everything that was measured or calculated. One point is to show that you can sort through all of your work and extract the main quantitative goals. Remember this point for future lab reports.

Significant Figures

The uncertainty dictates the number of significant figures that should be displayed for a result. First determine how many significant figures to keep for the uncertainty. Round off the uncertainty to one digit unless the first digit is a one or a two, in which case round off to two digits. The best value should then be rounded off, if necessary, to end in the same decimal place as the uncertainty. Of course, both must be expressed using the same units. Trailing zeroes may be needed to achieve the appropriate number of significant figures. Leading zeroes do not count.

Use the correct number of significant figures when expressing numbers as data or as results. Keep all digits in calculations, and when a result is achieved then apply the rules for determining how many significant figures to keep for the results.

(However, it is traditional not to keep more than two significant figures for *percentages*.)

Example 1-4: The current in a wire is measured to be 10.423 mA. Its uncertainty is estimated to be 0.59 mA. Express this result appropriately.

The uncertainty is dealt with first. Its first nonzero digit is a 5, which is not a one or a two, so it is rounded up to 0.6 mA. Both are expressed in mA, so no conversion is necessary. The uncertainty now ends in the tenths place, so the best value must also be rounded off to the nearest tenth. Thus, the result of this current measurement would be appropriately expressed as 10.4 ± 0.6 mA.

---

* * * * * * * * * What to Turn In * * * * * * * * * *

The Report: Lab reports should always include a descriptive title, the date on which the experiment was performed, and the first and last names of all partners. The report should be neat, well-organized, complete, legible, and written in complete sentences. Each student must individually complete and turn in a unique report. Plagiarism is a very serious problem that is not likely to be tolerated – it may result in a zero for the report, an automatic F in the course, or further disciplinary action.

Plagiarism

Copying the work of another and passing it off as your own constitutes plagiarism. This includes copying the lab report of another student, even if it is just part of it – even if it is just one phrase! It also includes copying information from a website or book. Do not plagiarize the work of others. Complete your work individually, in your own words. Prepare your work with your own unique style. Even calculations should reflect your own unique formatting, organization, and problem-solving approach. The only thing you should have in common with your partners is your data. If you receive help from another student, just seek minimal help where you are stuck and complete your report in your own unique way. If you provide help to another student, do not show your report to that student in order to prevent parts of your report from being copied.

If you cite information found in a textbook or online, attach a list of references to the end of your report and include a bracket and reference number after each part of your report that refers to the reference, using the format [#]. Any copied phrases should be in quotation marks. However, it is best to keep citations to a bare minimum, since the instructor is mostly concerned with assessing what you think, not what you can look up. Failure to cite any and all references constitutes an instance of plagiarism, which is not to be taken lightly.

The report for Experiment 1 should include the following sections:
- ✓ Data Table. The original data should be maintained in the laboratory notebook. Copy and organize this data into a table for inclusion with the report.
- ✓ Analysis. Computations should be well-organized, legible, and easy to follow. Begin with a symbolic formula, and show all steps of the calculations leading up to the results.
- ✓ Tabulation of Results. This is the table of main results for the experiment.
- ✓ Sources of Error. Type a paragraph discussing sources of error in detail.

Sources of Error

Identify which sources of error you feel are most significant for this experiment and discuss each in a separate short paragraph. Do not simply state the error, but describe its origin and how it affects the results in detail.

Do *not* use the words "human error" even if the error is human in origin because this is too vague. Be specific – equipment problems, e.g., is quite vague. Look for significant sources (e.g. round-off error is generally quite negligible compared to other errors) and refer to an inherent problem, *not* a mistake that could have reasonably been avoided (e.g. measuring the length to the top of the cylinder instead of its center of mass, reading the protractor incorrectly).

Adopt a positive, yet unbiased, outlook and a scientific tone. Do not write in the second person (e.g. "you," "your") or first person singular (e.g. "I," "me," "my").

In rare cases, you might be able to pinpoint where in the theory, and how, you could account for the error, or have a bright idea for a simple, effective way to reduce the error.

Research: Choose a research topic. Find one journal article or textbook (other than the required course textbook) at a suitable reading level that relates to your research project. Xerox no more than five pages that are relevant for your research topic. Attach a cover sheet to the xerox pages to submit as the first component of your research project (you may wish to keep a second copy for yourself). The cover sheet should include the subject of your research project, a brief description of your project, and a reference to the journal article or textbook in an appropriate bibliographic style.

References

List references in a separate section at the end of the report, entitled, "References" (not "Bibliography"). Anywhere in your report that you paraphrase the work of another, include a direct quote, or provide information from a source, you *must* cite your reference by enclosing the number of the reference in brackets (e.g. [4]) immediately following the paraphrase, quote, or referenced information. In the References section, list your references numerically, so that the numbers of your references match your citations.

Journal article references should be provided according to the following style, where all of the text is normal, except that the volume number appears in bold:

1. R. D. Young, "Physics of the Quark Model,"Am. J. Phys. **41**, 472 (1973).

The list of authors comes first, using the author's initials and last name, where a space separates the initials and a period follows each initial. For two authors, write "and" between the names, and for more authors, separate the names by commas and include "and" before the last name. A comma follows the list of authors, then comes the articles's title enclosed in quotation marks. A comma follows the title, before the closing quotation mark, then comes the abbreviation for the journal (which you can find online – common physics journal abbreviations and other useful style guidelines can be found under Submission Guidelines under Authors under Journals at www.aps.org) in normal text (not italicized). The volume number of the journal appears in bold. There is no comma immediately preceding the volume number. There may also be a letter for the journal, as in Phys. Rev. D**32**. A comma follows the volume number, then comes the number of the first page of the article. The year of publication is enclosed in parentheses, with no comma between the page and year. A period ends the reference at the end.

Book references should have the title of the work italicized as follows:

2. J. B. Marion, *Classical Dynamics of Particles and Systems*, 2nd. Ed. (Academic Press, New York, 1970), pp. 100-1.

Note the differences between the journal and book references. The title of an article appears in quotes, while a book title is italicized. The publisher information appears in parentheses for the book, and information is ordered differently. The last element of the book reference should refer to the chapter, section, or pages referenced.

It is important to adhere to the conventional style guidelines for physics references. Note that different disciplines follow different formats for citations and references, so you should be prepared for the style guidelines in other courses to differ somewhat from the style guidelines provided here.

Prelab Exercises for Experiment 2: Complete the following prelab exercises prior to the next lab.

1. Conceptually, describe how velocity relates to position.
2. Conceptually, explain what acceleration is.
3. Conceptually, explain the significance of the units of acceleration.
4. Mathematically, the formulas for average velocity and instantaneous velocity are very similar. What is the distinction? Under what conditions are they the same or different?

5. What common measuring devices could you use to help you determine the velocity of a running dog?

6. Describe, specifically, how the same measuring devices (although the number of measuring devices may be different) could be used to measure the acceleration of a dog from rest.

Experiment 2: Position, Velocity, and Acceleration

Introduction: It is rather straightforward to measure the velocity – a combination of speed and direction – of an object with uniform velocity (i.e. constant speed in a straight line). For example, if a car travels at a constant speed to the west and winds up 30 miles from its starting position half an hour later, it is easy to determine that its velocity was 60 mph to the west for the trip. However, if the speed changes during the course of the trip, determination of the velocity is not so simple. First of all, there is not a single velocity to speak of, since it varies over the course of the trip. The average velocity, based on the net displacement and elapsed time, does provide a useful measure, but often it is very helpful to determine the instantaneous velocity – i.e. the velocity at a specified time – or a table of instantaneous velocities.

The challenge in measuring the instantaneous velocity of an accelerating object is that the object is only traveling the speed in question for a very brief (infinitesimal!) moment of time. Thus, if measurements are made over a finite time interval, they will reflect that the velocity has changed during the elapsed time of the measurements. Nonetheless, a precise determination of the instantaneous velocity of an accelerating object can be obtained by computing the average velocity over a narrow window of time (centered about the particular time of interest, of course). The shorter the time interval, the more precise the instantaneous velocity.

Similar arguments can be made for average and instantaneous acceleration. Don't worry if this all sounds rather vague – it was intended to! The idea of this lab is to learn by discovery, specifically, the difference between the average and instantaneous values.

Textbook Reading: The textbook does discuss average velocity, instantaneous velocity, average acceleration, and instantaneous acceleration. Reading these sections will help aid in understanding this experiment. Pay special attention to the limit as Δt approaches zero in some of the definitions, which relates to theoretically gaining precision for shorter time intervals, as remarked above.

Objectives: Experimentally determine how position, velocity, and acceleration are related to each other, and learn the distinction between average and instantaneous values.

Apparatus: track, car, bar/picket fence attachment, photogates with mounting bracket and computer interface (or built-in digital readout), wooden block, level, vernier calipers, meterstick. A protractor or angle indicator may substitute for the vernier calipers and meterstick.

Procedures: The procedures for this lab have more procedural detail than normal. For one, this will serve as an example of what a detailed set of procedures looks like. Also, since there are multiple labs that use a track and car, the procedures offer more detailed instructions for this and related labs so that a variety of different measurements will be employed with the same equipment. As the course progresses, more independence will be expected from the student, and more creative inquiry-based opportunities will arise.

Level the track, via the leveling screw, if available, or by placing sheets of paper under one side. Position the car on the track and note how the wheels roll along the grooves. Take care during this and future experiments that the wheels are correctly positioned. Place the car in various positions to sample the overall levelness of the track, re-leveling the track if necessary to make it as level, on average, as feasible. Remember to begin with this step for future experiments that use the track and car.

Note

> If the photogate head is not already attached to the mounting bracket – as it should be – ensure that the proper length screw is used. For example, Pasco photogate heads use a 1.2-cm screw, while other Pasco parts use 1.5-cm screws; if one of the 1.5-cm screws is used, it will damage electrical parts inside.

Attach the mounting bracket to the track by sliding the large nut into the slot at the edge of the track (the side nearest to the computer interface). The photogates should not be too close to either end of the track. Gently roll the car along the track to check that no wires or other obstacles impede its motion.

The photogate measures the time at which each bar first triggers the photogate, and also the time interval during which each bar triggers the photogate. A computer interface uses the width of each bar (there is a default, which will likely be incorrect if the user does not check it) to compute the average velocity of each bar. For shorter bars, and hence shorter time intervals, the computed velocity is theoretically a better approximation to the instantaneous velocity. However, if the measured time interval is too short, there may be more experimental error in the computed velocity. The sampling rate can be adjusted in the software program to appropriate frequencies. If not using a computer interface, the instantaneous velocities must be computed manually.

Examine the five pattern picket fence, or equivalent bar. The five pattern picket fence includes two picket fences and three bars. Choose one of the three bars and record its band width. Place the five pattern picket fence atop the car (the five pattern picket fence fits snugly into a slot along the edge of the cabin of the car without fastening). Record the mass of the car with the five pattern picket fence.

If using a computer interface, turn it on and open the appropriate software program. Set the options for two photogates with bars (not picket fences). Enter the band width into the software program; you may need to do this separately for both photogates. If necessary, add a table to display the data. Select which quantities are to be displayed – only the elapsed time between the gates will be used for this experiment. For now, do not use any velocity or acceleration data that the program may offer to display. Ensure that you record elapsed gate-to-gate time, and not other times (such as the time during which one bar triggers one of the gates).

Note

> It is imperative that the correct band width be input into the software program. Otherwise, all of the measured velocities will be incorrect. Remember to do this in future experiments where a computer interface is used in conjunction with a photogate.

Adjust the height of the photogates such that the desired bar (the one for which you measured the band width) triggers each photogate as the car passes through it. Ensure that the entire length of the bar triggers the photogate properly by comparing the duration of the flashing red light on each photogate head to the time that the bar should be passing through (it is possible for a bar to trigger correctly for a moment, and then incorrectly trigger thereafter, if there is a slight positioning problem). Adjust each photogate head until it is perpendicular to the bar.

Develop a method for reliably reproducing a particular velocity on the level track (e.g. try using a built-in spring launcher). Position the photogates such that a relatively uniform velocity is expected passing through the two gates. Attempt a few trial runs prior to taking official data. The car should roll smoothly, with a minimum of vertical and side-to-side vibrations. Practice enabling and disabling data recording.

Record the distance between the photogates and the elapsed time between the two photogates directly into your laboratory notebook, as with all measurements made during lab. Repeat the elapsed time measurement a few times so that an average and standard deviation may be computed later. Change the distance between the photogates and record this new distance and a few more elapsed times. You will need a handful of varied distances between the photogates, in all.

Change the data table to display instantaneous velocity through the second photogate in addition to the elapsed time between the gates. Do not use any acceleration data that the program may offer to display.

Incline the track. Choose a starting position and take care to be able to reliably recreate the same starting position for each run, and to be able to reliably release the car from rest each time. Practice this a few times. Place one photogate near the car so that it is triggered just after release, but not before; this photogate will remain fixed. Vary the position of the other photogate. Gather a handful of data points for elapsed time and instantaneous velocity through the lower photogate, and record the distance. Check to see that the final velocity and elapsed time increase when the distance between the photogates increases; otherwise, there is a major problem to correct before leaving lab.

These were not the best experimental methods for using this equipment, but these are intuitive ways to go about measuring velocity. Improvements in the experimental technique will be made in future experiments.

Data Check: Prior to leaving lab, the following data should be gathered:
- ✓ a handful of distances between photogates, and a few elapsed times for each (level track)
- ✓ a handful of distances between photogates, and a few elapsed times and final velocities for each (inclined track)

Analysis: Plot the distance between the photogates as a function of the elapsed time for the level track. Determine what this graph represents physically, including the slope and y-intercept.

For the inclined track, plot the distance between photogates as a function of the elapsed time and also plot the final velocity as a function of the elapsed time – all on separate plots. Determine what these graphs represent physically, including the slopes and y-intercepts.

Plotting Data

Copy the data into a spreadsheet program such as Microsoft Excel. Highlight the data and insert a plot. Choose to make an xy scatter plot (a common mistake is to choose a line plot). Rather than assume that the first column corresponds to x or y, check this to be sure which is which, and correct it if necessary. Make the plot large, and adjust the limits of the axes to spread the data points out. Each axis needs a label (e.g. time) with units (in this case, seconds) enclosed in parentheses, as in "$t(s)$" – but without the quotes, of course. It is conventional to italicize the labels, but not the units. Include a descriptive caption, numbered if there are multiple plots. Horizontal and/or vertical gridlines should not be random, but should be helpful for reading points from the graph, if included; otherwise, tick marks, for example, need to achieve this effectively. Delete the legend when only one data set is shown in the graph.

For a data set that is predicted to be linear, add a trendline and display the equation on the graph. The value of R^2 is useful for determining how linear the data set is. For data sets that are predicted to form non-linear relationships, such as parabolas or circles, do not add a trendline (unless the data has been linearized, a technique that we will learn later). Remember to check these details for future reports that involve graphs.

Results: Make a table of the main quantitative results for this experiment.

$*$ $*$ $*$ $*$ $*$ $*$ $*$ $*$ $*$ **What to Turn In** $*$ $*$ $*$ $*$ $*$ $*$ $*$ $*$ $*$ $*$

The Report: The report for Experiment 2 should include the following sections:
- ✓ Data Table. Copy the measurements that you recorded into your notebook for this lab and organize them into a table.
- ✓ Analysis. This includes the graphs and any computations.
- ✓ Tabulation of Results. This is your table of the main results for the experiment.
- ✓ Conclusions. Type conclusions for this experiment.

Conclusions

The conclusions should be stated in a single paragraph. Begin by briefly summarizing, in one sentence or as part of a sentence, what was done experimentally. State the result(s) for this experiment. Draw conclusions, supporting any claims. For example, for which graphs does the theory predict a straight line? For those, do the data points support a linear relationship? What number can you refer to that would support this claim?

Avoid the word "prove," as it is incredibly strong – none of our labs "prove" anything because you can't "prove" something in a single lab. The spirit of science requires testing and retesting, and other researchers testing, etc. Make statements and support them by quoting numbers, describing patterns, etc. For example, you can say that you measured gravitational acceleration to be 9.2 ± 0.4 m/s^2, which differs with the accepted value by 1.5σ. Rather than say if this is good or bad, you can leave this judgment up to the reader. (Of course, this was not an example of a complete conclusion, but one example sentence showing one way to quantify agreement.)

As with all written sections of the report (so, remember this for future labs), adopt a positive, yet unbiased, outlook and a scientific tone. Do not write in the second person (e.g. "you," "your") or first person singular (e.g. "I," "me," "my"). Also, do not begin the conclusions by explicitly stating, "In conclusion."

Research: Type the introduction to your research project.

> **Introduction**
>
> The introduction will introduce the topic, provide some background on the problem and reference work that has been done in the past, define the problem that the project will solve, explain the need for numerical work, and describe what numerical work will be done and how it will solve the problem. Do not simply address each of these issues in order. The text should flow naturally – so there should not be any abrupt changes. The introduction should get the reader interested in your research. The introduction should *not* duplicate the Theory, Algorithm, Discussion, or other sections of your research paper. If you are looking for additional things to include in the introduction, you may compare your project to other methods, for example. The introduction should consist of a few paragraphs. Browse through some physics journal articles to develop a better idea of what an introduction should look like. The second half of the last paragraph of the introduction will provide a brief outline of the paper.

Prelab Exercises for Experiment 3: Complete the following prelab exercises prior to the next lab.

1. Each situation described below consists of two cars rolling down an inclined track. In each case, all factors are the same except for one. For each case, decide whether or not the two cars are predicted to have the same (or approximately the same, to within 1% or so) acceleration. If not, predict by what factor the accelerations should differ (e.g. should one be double the other?). For each case, explain the reasoning behind your answer in terms of physics concepts or by discussing a related equation.

   a. Two cars with identical mass roll down a 30° incline from rest. One car travels 1 m, while the other car travels 2 m.

   b. Two cars roll 1 m down a 30° incline from rest. One car has a mass of 250 g, while the other car has a mass of 500 g.

   c. Two cars with identical mass roll 1 m down a 30° incline. One car starts from rest, while the other has an initial velocity of 5 m/s.

   d. Two cars with identical mass roll 1 m along a 30° incline. One car rolls down the incline, while the other rolls up (obviously, not starting from rest).

   e. Two cars with identical mass roll 1 m down an incline from rest. One incline is tilted 30° relative to the horizontal, while the other is tiled 60° relative to the horizontal.

2. A 250-g car rolls 1 m up a 30° incline, then rolls 1 m back down (obviously, it did not start from rest). Take the positive coordinate to be up the incline. What is the car's acceleration:

   a. on the way up the incline?

   b. at its topmost position, just before rolling back down?

   c. on the way back down the incline?

Experiment 3: Uniform Acceleration

Introduction: A uniformly accelerated object undergoes constant acceleration – in 1D, it travels in a straight line and gains (or loses) speed at a constant rate. An object falling straight down (or up – and, yes, an object that is freely *falling* can move *up* during part or all of its trip; *all* in the case of escape velocity) near the surface of a planet experiences approximately uniform acceleration, provided that its change in altitude is small compared to the radius of the planet and neglecting other sources of error that are negligible for many situations. Similarly, an object sliding or rolling down (or up) an inclined plane experiences 1D uniform acceleration.

The advantage of utilizing an inclined plane, compared to vertical free fall, in lab to investigate uniform acceleration is that an object sliding or rolling along the incline will gain speed less rapidly (or lose speed less rapidly if heading up the incline), thereby reducing the relative error in the measured instantaneous velocities. There is also an advantage of using a car that rolls along the incline instead of an object, such as a block, that would slide along the incline. Think about this point, as it will come up again later.

Textbook Reading: Read the section in the textbook on uniform acceleration. Also read about Newton's second law and its applications – in particular, the motion of an object sliding along a frictionless incline. It will be useful to know the set of equations that apply to uniformly accelerated motion, and the equation for the acceleration of an object sliding along a frictionless incline.

Objective: Investigate the factors affecting the acceleration of a car down an inclined plane.

Apparatus: track, car, added masses, picket fence attachment, photogate with mounting bracket and computer interface (or built-in digital readout), wooden block, level, vernier calipers, meterstick.

Procedures: Examine the five pattern picket fence, or equivalent picket fence. Choose one of the two picket fences and record its band spacing (measured from the left edge of one bar to the left edge of the next bar). Notice that the band spacing is longer than the width of a bar. Also record the number of bars on the picket fence. Place the five pattern picket fence atop the car. Record the mass of the car with the picket fence.

If using a computer interface, turn it on and open the appropriate software program. Set the options for a single photogate with picket fence. Enter the band spacing into the software program. If necessary, add a table to display the data. Select which quantities are to be displayed – only the instantaneous velocity is needed. Do not use any acceleration data that the program may offer to display.

Note

It is imperative that the band spacing, and not the band width, be input into the software program when using a picket fence. Otherwise, all of the measured velocities will be incorrect. Remember how the band spacing and band width are defined and when each is needed for future experiments.

Adjust the height of the photogate such that the desired picket fence (the one for which you measured the band spacing) triggers the photogate as the car passes through it. Ensure that each bar of the same picket fence triggers the photogate properly by comparing the frequency of the flashing red light on the photogate head to the frequency of the bars of the picket fence (it is possible for a picket fence to trigger correctly for a few bars, and then incorrectly trigger thereafter, if there is a slight positioning problem). Adjust the photogate head until it is perpendicular to the picket fence.

Measure the height of the wooden block at a few different positions using vernier calipers. Incline the track by placing the wooden block beneath the pair of feet near the 100-cm mark, with the measured side of the block serving as the height. Measure the distance from the vertex made by the bottom of the leveling foot (or edge of the track if there is no foot) and the table to the position where the pair of feet are in contact with the block. This distance will be measured from the far edge of the leveling foot to the edge of the pair of feet nearest to the leveling foot. Do not simply read the numbers of the scale along the track: Observe that the zero of this scale does not correspond to the edge of the track. Estimate the uncertainty of the measurement of this length, and record this value with appropriate units.

Release the car from rest such that it develops a nonzero velocity prior to triggering the photogate, yet if the velocities are too great there will more error in the measured time intervals. Attempt a few trial runs prior to taking official data. The car should roll smoothly, with a minimum of vertical and side-to-side vibrations. Practice enabling and disabling data recording. Compare the number of data points to the number of bars of the picket fence. Do not be alarmed if there are one or two more bars than data points, but if the discrepancy is greater then there is a problem to be resolved. After a smooth run, copy the table of velocities and times to Excel, and save the data in Excel. Also, copy the table in your lab journal.

Collect additional data as needed in order to achieve the objectives for this experiment, several examples of which were provided in the prelab exercises.

Data Check: Prior to leaving lab, the following data should be gathered:
- ✓ number of bars of picket fence, band spacing of picket fence
- ✓ few heights of wooden block
- ✓ length of incline from vertex to wooden block, including uncertainty
- ✓ tables of velocities and times for various factors

Analysis: Compute the angle of the incline. Use this to compute the theoretical acceleration of the car down the incline. Make a linear plot from the data and determine the experimental acceleration in terms of the slope. Determine the uncertainty in the slope from a linear regression or by drawing maximum and minimum slopes, and establish the uncertainty in the experimental acceleration based on this.

Uncertainty in the Slope

The uncertainty in the slope of a linear plot can be found by performing a linear regression, as outlined in Appendix A, or from a less rigorous method of minimum and maximum slopes. The linear regression method is an algorithm that can be applied by writing a computer program or using a spreadsheet program such as Excel. After making a linear regression program once, it will be very convenient to reuse the same program for every experiment where a linear plot is made, which is quite often.

Alternatively, the uncertainty in the slope can be found as follows. Print out the graph. Draw reasonable minimum and maximum slopes that reflect the scatter of the data points; this requires some judgment and reason. Find the slopes of these two lines. Half of the difference between these slopes equals the uncertainty in the slope of the best-fit line.

Results: Make a table of the main quantitative results for this experiment. State results in the format $x \pm \sigma_x$. Include the percent error or difference.

Percent Error/Difference

The percent error between a computed value and an accepted value is calculated as

$$\frac{|\text{computed} - \text{accepted}|}{\text{accepted}} 100\%$$

If instead two values are compared where neither value has more scientific acceptance than the other, a percent difference can be found. In this case, the denominator equals either the average or the smaller of the two values (these being the two popular conventions).

---

* * * * * * * * * * What to Turn In * * * * * * * * * * *

The Report: The report for Experiment 3 should include the following sections:
- ✓ Data Table.
- ✓ Analysis.
- ✓ Tabulation of Results.
- ✓ Discussion of Results. Type a discussion of the results for this experiment. As a side point, examine why friction is negligible for the car rolling down the incline.

Discussion of Results

Organize your discussion in the format of a classic five paragraph essay including an introductory paragraph, a three-paragraph body, and conclusion. Your thesis should be geared toward the result of the main objective for the experiment.

The introductory paragraph serves as a lead-in and brief overview. Save your conclusions for the final paragraph – do *not* state them in the introduction or body. The body should gradually build up to the point where you are ready to draw conclusions naturally in the final paragraph. Your analysis should flow logically, and all claims should be supported immediately – by quoting results, referring to an equation, etc.

Your discussion should reflect your ability to analyze data and results. Examine whether or not a data set forms a specific pattern, and, if so, describe the pattern. Decide whether or not the deviations are significant – i.e. are the deviations smaller or greater than inherent experimental errors? If declaring deviations to be significant, briefly refer to a specific source of error that is probably a primary factor (but not if it is inconsistent with the discrepancy). You can quantify deviations – e.g. standard deviation or percentage. You can reference a plot (e.g. by referring to Fig. 1) to support if data obey a linear relationship; Excel provides a number to help support this. You can also quantify the agreement between theory and experiment by stating the discrepancy in terms of σ (described in the write-up to Experiment 4).

Research: Develop the formalism for your research project.

Formalism

In paragraph form, with equations embedded using Microsoft Word's equation editor, introduce the reader to the theory, including descriptive text and discussion of the concepts and strategy in addition to mathematical equations. Derive equations as part of your discussion, clearly explaining the steps. Your goal is for a beginning student to be able to read your Formalism section and be able to follow it. The formalism section of your research paper should accomplish the theoretical goals of your project. (Illustrations for your research project will be assigned in the next lab.)

Prelab Exercises for Experiment 4: Complete the following prelab exercises prior to the next lab.

1. Vector $\vec{A}$ has a magnitude of 123 N and y-component of 45 N. What are the possible directions of $\vec{A}$? (Note: Measure positive angles counterclockwise from the x-axis.)

2. Vector $\vec{A}$ has a magnitude of 531 N and direction of 123°. Vector $\vec{C}$ has a magnitude of 642 N and direction of 246°. Find the magnitude and direction of $\vec{B}$, where $\vec{B} = \vec{C} - \vec{A}$. (The magnitude of $\vec{B}$ is *not* 111 N!)

Experiment 4: Vector Addition

Introduction: Some measurable quantities involve not only a magnitude (essentially, how much), but also have an inherent direction. Such quantities are termed *vectors*. Other physical quantities feature just a magnitude, lacking a directional attribute. These quantities are called *scalars*.

For example, consider the two seemingly similar quantities, mass and weight. These two distinct quantities have fundamentally different directions, yet they are often confused and incorrectly used interchangeably, largely because it turns out that the magnitude of an object's weight in the presence of a gravitational field is proportional to the object's mass: $W = mg$. In physics, however, there are many situations where it is conceptually crucial to recognize the fundamental differences between these two quantities.

Weight is the gravitational pull that one mass (usually, astronomical, like a planet or star) exerts on another mass. For example, your weight is the force that the earth exerts on you. The force that the earth exerts on an object is directed toward the center of mass of the earth. Thus, weight acts in one particular direction. An object on the surface of a planet experiences weight as a downward pull. If you drop a textbook out of the window, it will definitely fall down.

Mass is a measure of how much inertia an object has. Put another way, mass is the 'reluctance' of an object to accelerate (using a little personification – surely, the object does not have a choice in the matter). To see this, try pushing various objects. For example, it is pretty easy to accelerate a pencil, the same push will accelerate a pencil much more than a textbook, and much more force is required to accelerate a desk. The desk has more mass, the pencil has much less mass. Mass is not a reluctance to *move*, though, but to be *accelerated*. The distinction is that an object in motion has a natural tendency – inertia – to maintain constant velocity. You can catch and stop a runaway skateboard, but you need to be superhuman to stop a runaway car, which has much more mass.

Mass, unlike weight, does not have an associated direction. To see this, try pushing a desk in various horizontal directions. You should observe that it is equally difficult to accelerate a desk to the east as it is to the north. Mass is a scalar, while weight is a vector.

Textbook Reading: Read the sections of the textbook on scalars, vectors, components, graphical vector addition, using trigonometry to add vectors, and vector subtraction.

Objective: Design and conduct experiments to test the theory of vector addition for:
- finding the components given the magnitude and direction of a vector
- computing the magnitude and direction of the resultant of two or more vectors
- the effect of subtracting one vector from another

Apparatus: force table, pulleys with clamps, ring, thread, scissors, mass holders, set of masses, level. Alternatively, this experiment can be performed using metersticks (or tape measures) and protractors.

Procedures: Get acquainted with the features of the force table, then set the force table up for the experiment. Observe that the screw prevents the ring from falling when the system is not in equilibrium. As the system approaches equilibrium, through adjustment of the unknown load and angle, the screw should be lowered out of the way to allow for some fine-tuning. The force table may have adjustable pulleys – i.e. they can be raised while searching for equilibrium, and then lowered to measure the angle more precisely. Level the force table. Note that the angles should be recorded to the nearest tenth of a degree.

Caution

> Note the maximum load that the force table can accommodate. Be sure not to exceed this load during the experiment.

Notes: Keep the following notes in mind when you conduct the experiments described in the next paragraph. The force table allows you to experimentally locate the position and magnitude of the equilibrant (the balancing force). A small increment of mass may be added or subtracted from the load without noticeably disturbing equilibrium. That is, there is a range of loads pertaining to equilibrium. Add just enough mass to disturb equilibrium to find the upper bound, and subtract mass to find the lower bound. Record the mass of the minimum and maximum loads resulting in equilibrium. The average load is the best value for the mass, and one-half of the difference is the uncertainty in the mass of the load. Adjust the load to the average, and then locate the minimum and maximum angle. Similarly, record the best value for the angle and the uncertainty in the angle. Also record the loads and angles of the given values.

Conduct experiments to determine the information described below, where the magnitude of $\vec{A}$ is twice the magnitude of $\vec{B}$ and $\vec{A}$ has a direction of 155° and $\vec{B}$ has a direction of 220° (where the angles are measured counterclockwise from the positive x-axis):

1. Experimentally determine the components of $\vec{A}$.
2. Experimentally determine the magnitude and direction of the resultant of $\vec{A}$ and $\vec{B}$.
3. Experimentally determine the magnitude and direction of $\vec{C}$, where $\vec{C} = \vec{B} - \vec{A}$.
4. Conduct a simple experiment to determine whether or not it was necessary to level the force table in the beginning of the experiment.

Data Check: Prior to leaving lab, the following data should be gathered:
- ✓ the magnitudes and directions of given vectors ($\vec{A}$ and $\vec{B}$)
- ✓ the components of $\vec{A}$
- ✓ the magnitude and direction of the resultant of $\vec{A}$ and $\vec{B}$
- ✓ the magnitude and direction of $\vec{C}$
- ✓ a description of the experiment to see whether or not leveling was necessary, and data from this experiment

Analysis: Calculate the theoretical unknown for each experiment and compare with the experimentally measured unknown and its uncertainty.

Results: Make a table of the main quantitative results for this experiment. State results in the format $x \pm \sigma_x$. Include relevant percent errors or differences and state the relative discrepancies.

Relative Discrepancy

The discrepancy is the difference between two determined values; often, the difference between an experimentally determined value and an accepted value. The relative discrepancy expresses the difference in terms of the uncertainty. For example, comparing 9.81 m/s² to 9.62 ± 0.15 m/s², the discrepancy 0.19 m/s² is 1.3 times larger than the uncertainty 0.15 m/s², so this is a 1.3-σ discrepancy. This is a useful way of establishing the level of agreement between two results. Without first establishing the uncertainty, it is impossible to determine whether or not two results agree. For example, it might seem that a measurement of 99.65 kg agrees quite well with 97.23 kg, since they differ by only 2.5%, but if the uncertainty is 0.03 kg, then there is no agreement at all, since the results differ by 81σ! On the other hand, 7.2 V agrees with 8.1 V within one standard deviation if the uncertainty is 0.9 V or higher. We strive for small uncertainties combined with excellent agreement.

* * * * * * * * * **What to Turn In** * * * * * * * * * *

The Report: The report for Experiment 4 should include the following sections:
- ✓ Abstract. Type an abstract for this experiment.
- ✓ Data Table.
- ✓ Analysis.
- ✓ Tabulation of Results.
- ✓ Sources of Error.

Abstract

The abstract is a concise (about 2-3 sentences only) synopsis, with a positive scientific tone, of: what was done, a statement of numerical values ± uncertainties for key results, a statement of the level of agreement (% and/or discrepancy in terms of σ), and a brief statement of key conclusions (or a plausible source of error consistent with results if there is a significant discrepancy).

The abstract is useful in scientific publications for readers to quickly sort through a myriad of journal articles to find just the few that are relevant for their research interests. Thus, it is important to be very concise. Combine multiple ideas into a single sentence, and do not waste words. Do not devote an entire sentence to what was done experimentally (just mention it briefly in passing).

Be specific. Do not explicitly refer to the experiment or the purpose. Be results-oriented (do not describe calculations or graphs). State results before stating conclusions. State results (what you found) instead of objectives (what you set out to do).

Sample abstract: The energy stored in a compressed spring with constant 15 N / cm was found to vary quadratically with its displacement from equilibrium by utilizing its potential energy to launch a PasCar up an inclined track. The spring constant was found to be 14.4 ± 0.4 N / cm, with the 1.5σ discrepancy attributable to energy lost to minute vertical and side-to-side vibrations of the PasCar's wheels.

Research: Draw and label professional-looking illustrations relevant for your research project.

Illustrations

Draw professional-looking, labeled diagrams. High-quality diagrams can be drawn with Microsoft Word (with some thought, effort, and patience), for example, but do not print with high quality in Paint and many other programs. In Word, turn the gridlines on or off as needed – you do not want to snap objects to gridlines or other objects when you want the freedom to place it exactly where you want, but other times you do want things to be evenly spaced or snap together. Remove the outlines of textboxes. Italicize symbols, but not units, and place vectors in boldface or draw an arrow over them using the equation editor. You can group objects together and rotate an entire group, or select multiple objects and align them, for example. If you explore the options, you may find that you can do much more than you may first suspect.

Prelab Exercises for Experiment 5: Complete the following prelab exercises prior to the next lab.

1. A girl throws a ball with an initial speed of 20 m/s from the roof of a building. The ball is 12 m above the ground when it leaves the girl's hand. How far does the ball travel horizontally before striking the ground if it is thrown:
   a. horizontally?
   b. at an angle of 30° above the horizontal?
   c. at an angle of 30° below the horizontal?

Experiment 5: Projectile Motion

Introduction: Very often for a launched projectile, such as a football pass or a launched missile, it is useful to predict where the projectile will land. The calculations – known as *ballistics* – can become very complicated, accounting for variation of gravitational acceleration with altitude, air resistance, and the rotation of the earth, for example. However, in many cases the approximate landing area may be computed to a high degree of accuracy by assuming that the projectile is in a state of free fall – i.e. the only force acting upon it is the downward pull of gravity – and assuming gravitational acceleration to be uniform.

Textbook Reading: Read the section of the textbook on projectile motion. The related equations and problem-solving strategy for projectile motion will be useful.

Objective: Measure the landing position of a launched ball to determine the velocity of the ball as it leaves the barrel of a projectile launcher, then change the launch angle and predict the new landing position.

Apparatus: projectile launcher, ballistic pendulum, barrel loader, steel ball, meterstick, ruler, landing pad, collection box, white paper, carbon paper, target, tape, level, C-channel clamp. An alternative to the projectile launcher is a track from which a steel ball can roll off, but this is more limited unless the launch angle can be varied in a way that the initial velocity is still known for the predictive part of the experiment (otherwise, the height can be changed, for example, instead of the launch angle).

Procedures: Choose a range setting for the projectile launcher. Clamp down the projectile launcher.

Caution

Do not look into the barrel of the launcher. Wear safety goggles. Do not place your hand or anything else in front of the launcher. Do not leave a ball in the barrel of the launcher unattended – occasionally, there are accidental launches. After loading a ball, keep the space clear – that everyone is out of the way, and knows that you are about to launch a ball so that they do not walk in the potential path of the ball – until after it is launched.

     With the space clear, use the barrel loader to load a ball into the barrel launcher. Launch the ball while the space is clear, and do not obstruct the launch path during the interim period.

Caution

Do not load the ball into the projectile launcher with your finger. Use the barrel loader to do this to prevent injury.

Place a sheet of carbon paper beneath the target and make a test mark on one corner to verify that you have these stacked together properly. Place a collection box behind the target. Also, place the target and carbon paper atop a landing pad to cushion the floor against a dent, which is sturdy enough for a mark to be made on the carbon paper.

Make distance measurements and use these to compute the initial velocity of the ball from the ball launcher (just after leaving the barrel).

Change the angle of the launch and use the initial velocity to predict where the ball will land. Now place the target in this predicted position. Do not load a ball until your instructor is ready to observe your inclined launch. Premature launches are not permitted.

Data Check: Prior to leaving lab, the following data should be gathered:
- ✓ distance measurements needed to compute the initial velocity
- ✓ measurements involved in predicting the target position for the inclined launch
- ✓ actual landing position of the ball for the inclined launch

Analysis: Organize clearly your complete calculation of where the ball should land for the inclined launch, beginning with your calculation of the initial velocity.

Results: Make a table of the main quantitative results for this experiment. Include the percent error or difference.

* * * * * * * * * * **What to Turn In** * * * * * * * * * * *

The Report: The report for Experiment 5 should include the following sections:
- ✓ Data Table.
- ✓ Analysis.
- ✓ Tabulation of Results.
- ✓ Experimental. Type an experimental section for this experiment.

Experimental

In a half-page (or little more) paragraph, describe the details of the experiment. An experimental section is different from a set of procedures. For one, an experimental section says what was done, whereas procedures say what to do. The experimental section is written in paragraph form, *not* as numbered steps. Also, the procedures of this manual serve as a guide, and omit important steps and details in an effort to harness independence from the experimenter, but the experimental section is written with a different purpose and so should include these steps and details. Try to think of what you were wondering about during the experiment that would have made your life easy had they been explained in the procedures.

In future labs, look ahead to see if you need to write an experimental section; if so, jot down notes in your notebook that will help you write this later.

Research: Write the algorithm for your research project. Add references and submit an updated References section.

Algorithm

The algorithm describes how your computer program or spreadsheet file performs the calculations. This section should first describe what the algorithm does and how it relates to your research project. It should then list user input, describe the body of the algorithm, and list the output.

If you looked up any algorithms, include a xerox copy of the algorithm and write the reference on the top of the first page. While you may look up an algorithm in words, you may not copy or use any subroutines or programs. You must write your own code or spreadsheet file from scratch; the algorithm serves as a guide for what to do.

You are not turning in the program itself, nor the code – just the algorithm. The algorithm is a descriptive outline that explains, specifically, what your program tells the computer to do. Describe any numerical routines, specifically outlining how each is implemented. The goal is that anybody who reads your algorithm can sit down and write a program to carry it out without any trouble (even if they do not know that same programming language that you used – since, again, you are not writing the code itself, but a description of what it does).

Prelab Exercises for Experiment 6: Complete the following prelab exercises prior to the next lab.

1. Find the derivation of the acceleration of the system of masses in the textbook for Atwood's machine, if available; if not available, begin with a free-body diagram for each mass and apply Newton's second law, following similar examples from the textbook, in order to derive this equation. The following questions pertain to this derivation. Call m_2 the greater mass and m_1 the lighter mass.

   a. The two masses exert a tension force on one another through the connecting cord. Which mass exerts the greater tension force on the other?

   b. How does your answer to the previous question (a) relate to the derivation?

   c. When one mass accelerates up, the other accelerates down. Where does this figure into the derivation?

   d. List some simplifying assumptions that are made in order to make the solution more manageable.

   e. Which mass experiences greater acceleration? Explain.

   f. In the limit that the two masses become equal, what happens to the acceleration?

   g. What would the acceleration of the masses be if the cord came untied?

Experiment 6: Gravitational Acceleration

Introduction: Atwood's machine consists of a cord that passes over a pulley, connecting to a mass at each end, as illustrated in Fig. 6-1. For similar reasons to those discussed in the Introduction to Experiment 3: Uniform Acceleration, since the acceleration of the system of masses for Atwood's machine is significantly reduced compared to gravitational acceleration, Atwood's machine comes with an advantage over a vertical free fall experiment for determining gravitational acceleration.

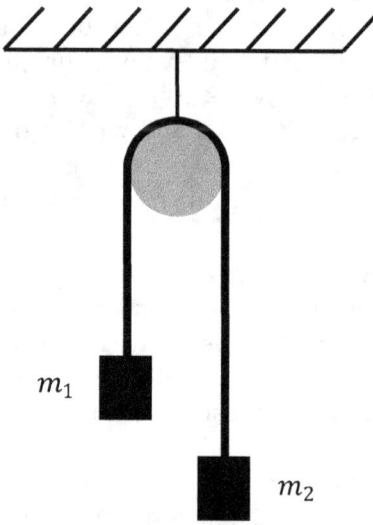

Fig. 6-1. The basic structure of Atwood's machine.

Textbook Reading: Read the sections in the textbook on Newton's second law and its applications, and closely follow any derivations or examples pertaining to Atwood's machine. The strategy for applying Newton's second law and the derived equation for the acceleration of the masses for Atwood's machine will be needed.

Objective: Use Atwood's machine to determine gravitational acceleration near earth's surface.

Apparatus: smart pulleys (including photogates and posts), table clamps, computer interface, set of slotted masses with hangers, thread, scissors, vernier calipers.

Procedures: Construct the Atwood's machine setup. If two pulleys are used, they should be aligned. The threads should smoothly pass through the grooves of the pulleys and not make contact with the clamps, table, or any other obstacles. Practice the technique of releasing the pulleys to ensure that the two sections of thread are vertical upon release and during the acceleration of the masses. Swaying should be kept to a minimum. Try to prevent the masses from colliding. If they do collide, this is not a source of error: Simply discard that data run, so that results are not affected (otherwise, it is not an *error*, but a *mistake*).

Verify that the smart pulley's photogate is properly triggered by the pulley's spokes. Set up the computer interface to use the smart pulley sensor. Measure the circumference of the pulley inside the groove where the thread lies and divide this length by the number of spokes. Verify that the preset constant in the data collection software program agrees with this incremental arc length, updating it if necessary.

Gather data for velocity as a function of time for several different pairs of masses that add up to the same value – e.g. 50 g. Do not keep extra data that the program collects that corresponds to one of the masses having struck the floor or the thread having slipped off the pulley, for example.

Repeat the data runs where all the masses are multiplied by three.

Data Check: Prior to leaving lab, the following data should be gathered:
- ✓ data table for velocity as a function of time for several pairs of masses, where the sum of the masses is constant
- ✓ similar data table where the sum of the masses is three times greater

Analysis: For each pair of masses, make a linear plot and compute the acceleration in terms of the slope, including its uncertainty.

Results: Make a table of the main quantitative results for this experiment. State results in the format $x \pm \sigma_x$. Include relevant percent errors or differences and state the relative discrepancies.

* * * * * * * * * *   **What to Turn In**   * * * * * * * * * *

The Report: The report for Experiment 6 should include the following sections:
- ✓ Abstract.
- ✓ Data Table.
- ✓ Analysis.
- ✓ Tabulation of Results.
- ✓ Discussion of Results.

Research: Begin implementing the algorithm. Include a note that describes what progress you have made on your algorithm thus far. Begin a discussion for your research project. This will be similar to a Discussion of Results, but geared toward your research project. One thing you can do now is describe how your numerical program is being employed. You will need to obtain numerical results in order to complete your discussion, so just write what you can with what you have thus far. If you have no results at this point, you can at least start to make some comparisons and begin other ingredients to your discussion, and then complete the discussion when you obtain results. Since the discussion is one of the heartier sections of the research paper, it is important to make some progress on it now.

Prelab Exercises for Experiment 7: Complete the following prelab exercises prior to the next lab.
1. Design an experiment to determine the coefficient of kinetic friction between two surfaces that uses some or all of the equipment listed under Apparatus for Experiment 7. The design should include:
   a. a labeled diagram of the experimental setup

    b. a list of equipment used

    c. a set of procedures that thoroughly describe the details of how the experiment will be conducted, including which quantities will be measured and how the measurements will be made

    d. a symbolic derivation, starting with a free-body diagram and Newton's second law, for the coefficient of kinetic friction

2. For the case of static friction, either explain how this result can be obtained as a simplification of the experiment described in answer to Exercise 1 and show how the derivation simplifies for the coefficient of static friction, or design a new experiment and repeat the steps of Exercise 1 as they apply to static friction.

Experiment 7: Friction

Introduction: Friction is a contact force between two surfaces in relative motion that resists, or at least limits, the relative acceleration of the two objects in contact. Friction may make the mathematical nature of problems somewhat more involved in physics problems, but life would be drastically different, but probably not for the better, without it. To appreciate this, watch the news after the next winter storm leaves the highways and roads icy.

A force of static friction acts to prevent or limit the acceleration of a stationary object that would otherwise occur. For example, any object placed on a frictionless incline will accelerate to the bottom, but a block placed on an incline with friction may remain at rest on the incline, depending upon whether or not the component of the block's weight along the incline exceeds the force of static friction. For an object in motion, a similar force of kinetic friction resists, or limits, the acceleration of the object. The reason for the different terms, static and kinetic, is that the coefficient of friction may vary, depending upon whether or not the object is at rest or in motion.

Textbook Reading: Read the sections of the textbook on the coefficient of friction, friction force, and application of Newton's second law to problems involving friction.

Objective: Determine the coefficients of static and kinetic friction between two surfaces.

Apparatus: track, friction carts, smart pulley (including photogate and post), table clamp, computer interface, wooden block, set of slotted masses with hangers, vernier calipers, level, meterstick, thread, scissors, scale. A suitable block that thread can easily be tied to can serve as a friction cart. It is possible, though less convenient, to use regular photogates triggered by a cart or block attachment instead of a smart pulley.

Procedures: Carry out the experiments you designed as a prelab exercise.

Data Check: Prior to leaving lab, the following data should be gathered:
- ✓ Make a list of what data you need in order to meet the objectives of your experiment. Record qualitative observations as well as quantitative data

Analysis: Compute the coefficients of static and kinetic friction for each pair of surfaces, including uncertainties.

Results: Make a table of the main quantitative results for this experiment. State results in the format $x \pm \sigma_x$. Include relevant percent errors or differences and state the relative discrepancies.

* * * * * * * * * *   What to Turn In   * * * * * * * * * * *

The Report: The report for Experiment 7 should include the following sections:
- ✓ Data Table.
- ✓ Analysis.
- ✓ Tabulation of Results.

✓ Sources of Error.
✓ Conclusions.

Research: Continue working on your algorithm. Make some preliminary graphs of your numerical data, based on what your program is able to do thus far. As always, describe each graph in the caption.

Prelab Exercises for Experiment 8: Complete the following prelab exercises prior to the next lab.
1. Which types of energy must be computed in order to determine if energy is conserved for a car rolling down an inclined track?
2. For each type of energy, provide a formula that will be useful for calculating the energy.
3. For each symbol in each formula, indicate if it is a known constant or describe what equipment can be used to measure it and how the measurement can be made.
4. If the track is curved rather than straight, will this cause an error in the results of the experiment? Explain.

Experiment 8: Conservation of Energy

Introduction: Conservation of energy is a powerful problem-solving technique because it is often path-independent. Consider, for example, an object sliding down a frictionless hill. The equations of uniform acceleration only apply if the hill has a constant incline; if the hill is instead curved, the acceleration is non-uniform, and mathematically it becomes a much more complicated matter to relate position, time, instantaneous velocity, and instantaneous acceleration. However, energy is conserved for the object as it slides down the frictionless hill, regardless of whether or not the hill is curved; and the strategy for conserving energy is equally simple to apply either way. Conservation of energy is, in general, very useful for relating positions to speeds.

Textbook Reading: Read the sections of the textbook on energy, potential energy, kinetic energy, and conservation of energy. As energy relates to work, it would also be wise to read the section on work. The equations for gravitational potential energy and kinetic energy and the strategy for conserving energy will be applied in this experiment.

Objective: Determine to what extent mechanical energy is conserved for a car rolling along an inclined track.

Apparatus: track, car, added masses, bar/picket fence attachment, two photogates attached to mounting brackets, computer interface, wooden block, vernier calipers, level, meterstick, scale.

Procedures: Set up two photogates and set up the computer interface to record the instantaneous velocity of the car through the two photogates; use a bar, *not* a picket fence, and enter the appropriate constant for the bar for both photogates.

Gather data that will be useful for determining whether or not energy is conserved for a car rolling along an inclined track. Measure any distances to the nearest tenth of a millimeter. Determine the uncertainties in your measurements. The uncertainty in a velocity can be determined after the experiment is otherwise over by making the track horizontal and switching to a picket fence: Roll the car through a few times to approximately recreate the desired velocity, and the standard deviation will give you something to work with. There may be additional factors to consider, but this is a starting point.

Data Check: Prior to leaving lab, the following data should be gathered:
- ✓ data necessary to determine if energy is conserved for a car rolling along an inclined track, including uncertainties

Analysis: Perform calculations with quantitative data to determine to what extent energy was conserved. Include uncertainties.

Results: Make a table of the main quantitative results for this experiment. State results in the format $x \pm \sigma_x$. Include relevant percent errors or differences and state the relative discrepancies.

❋ ❋ ❋ ❋ ❋ ❋ ❋ ❋ ❋ ❋   What to Turn In   ❋ ❋ ❋ ❋ ❋ ❋ ❋ ❋ ❋ ❋

The Report: The report for Experiment 8 should include the following sections:
- ✓ Abstract.
- ✓ Data Table.
- ✓ Analysis.
- ✓ Tabulation of Results.
- ✓ Sources of Error.

Research: Complete your numerical algorithm. Type an abstract for your research paper, based on what you have thus far. Complete your Discussion section, but save conclusions for a separate Conclusions section (assigned in the next lab).

Prelab Exercises for Experiment 9: Complete the following prelab exercises prior to the next lab.
1. Design an experiment to meet the various objectives outlined in Experiment 9, using some or all of the equipment listed under Apparatus. Include:
   a. a labeled diagram of the experimental setup
   b. a list of equipment used
   c. a set of procedures that thoroughly describe the details of how the experiment will be conducted, including which quantities will be measured and how the measurements will be made
   d. a description of how the needed analysis will be performed
2. How does the fact that momentum is a vector enter the calculations if the collision is 1D?

Experiment 9: Conservation of Momentum

Introduction: It is particularly useful to conserve momentum for collisions. There are two types of collisions: elastic and inelastic. Of the inelastic collisions, the completely inelastic collision is an important and common special case. Contrary to popular expectation, a collision for which objects bounce apart may not be elastic, and this counterintuitive problem is the source of some major mistakes in problems with collisions. In physics, elastic refers to a collision in which both momentum and mechanical energy are conserved, whereas mechanical energy is lost in an inelastic collision (or gained, as in the case of an inverse inelastic collision). A completely inelastic collision is one in which the objects stick together after colliding.

Textbook Reading: Read the sections of the textbook on momentum, conservation of momentum, and the various types of 1D collisions. The equation for momentum and the technique for conserving momentum for the different types of 1D collisions will be utilized.

Objective: Investigate when linear momentum and/or kinetic energy are conserved for various elastic and inelastic collisions.

Apparatus: track, cars that can collide approximately elastically or completely inelastically, added masses, bar/picket fence attachment, two photogates attached to mounting brackets, computer interface, vernier calipers, level, scale.

Procedures: Carry out the experiments you designed as a prelab exercise. Following are a few tips:
- Check that the track is level. Place a car on the track to check different portions of the track.
- Signs are crucial. The computer interface can measure speed, but you must record the directions manually.
- Strive for a smooth, clean collision, minimizing vertical and side-to-side vibrations.
- A car should not trigger a photogate during a collision, but should not have a large distance over which resistive forces can have a significant effect before triggering the photogate.

Data Check: Prior to leaving lab, the following data should be gathered:
- ✓ Make a list of what data you need in order to meet the objectives of your experiment. Record qualitative observations as well as quantitative data. Include uncertainties.

Analysis: Perform calculations with quantitative data that show whether or not linear momentum and/or kinetic energy were conserved for each collision. Include uncertainties.

Results: Make a table of the main quantitative results for this experiment. State results in the format $x \pm \sigma_x$. Include relevant percent errors or differences and state the relative discrepancies.

* * * * * * * * * * What to Turn In * * * * * * * * * *

The Report: The report for Experiment 9 should include the following sections:
- ✓ Data Table.
- ✓ Analysis.
- ✓ Tabulation of Results.
- ✓ Experimental.
- ✓ Conclusions.

Research: Check your algorithm for mistakes and check your numerical results for consistency, that they make sense, and that they agree with theoretical examples or behave appropriately in certain limits. Type a Conclusions section for your research paper.

Prelab Exercises for Experiment 10: Complete the following prelab exercises prior to the next lab.
1. Derive the symbolic equations described below, which will be needed during Experiment 10.
    a. Derive an equation that relates h_b to L and θ_m. This is a math problem based on the diagram, not involving physics concepts.
    b. Conserve energy for the upward swing of the ballistic pendulum (with ball) in order to derive an equation for v_b in terms of m_p, m_b, L, θ_m, and/or g.
    c. Conserve momentum for the collision at the bottom in order to derive an equation for v_p in terms of m_p, m_b, L, θ_m, and/or g.
    d. Apply the strategy of projectile motion to derive an equation for the horizontal distance R that the steel ball will travel before striking the floor in terms of m_p, m_b, L, θ_m, h_p and/or g.

Experiment 10: Conservation of Momentum and Energy

Introduction: The ballistic pendulum experiment involves launching a projectile that makes a completely inelastic collision with a pendulum, causing the pendulum to swing upward. It is a prime example where the techniques of conserving momentum and energy are combined, though applied separately, to make useful physical predictions.

A steel ball of mass m_p is launched horizontally with initial speed v_p from a projectile launcher from a height h_p. The steel ball immediately lands in a ballistic pendulum with mass m_b, as illustrated in Fig. 10-1. The length of the ballistic pendulum, L, is measured from the point of support to the center of mass of the ballistic pendulum (with ball). The center of mass is readily found by detaching the ballistic pendulum (with ball), and locating the approximate balancing position by finger. Note that p refers to the projectile (the ball), while b refers to the ballistic pendulum.

The steel ball collides with the ballistic pendulum in a completely inelastic collision. The speed of the ballistic pendulum (with ball) is v_b immediately after the collision. Following the collision, the ballistic pendulum (with ball) rises to a maximum angle θ_m, corresponding to a raised height h_b.

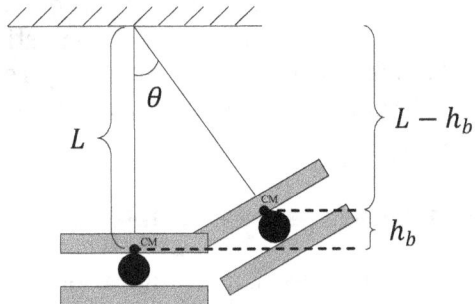

Fig. 10-1. The initial and final positions of a ballistic pendulum are shown.

Measurements of the mass of the ball, mass of the ballistic pendulum, length of the ballistic pendulum, and maximum angle can be used to compute the initial speed of the steel ball from a projectile launcher. The ballistic pendulum can then be removed. Where the steel ball will land on the floor can be predicted via calculation.

Textbook Reading: Review the sections on energy, momentum, conservation of energy, and conservation of momentum, and read any examples or sections on the ballistic pendulum. The equations for momentum, potential energy, and kinetic energy, along with the strategies for conserving energy and for conserving momentum, will be needed during this experiment. Also review the sections on projectile motion, as the equations and problem-solving strategy for projectile motion will also be needed during the experiment. This experiment requires extensive calculation during lab, so it will be well worthwhile to come to the experiment prepared.

Objective: Apply conservation of momentum, conservation of energy, and projectile motion problem-solving strategies to predict where a projectile launcher will launch a steel ball onto the floor from ballistic pendulum measurements.

Apparatus: projectile launcher, ballistic pendulum, barrel loader, steel ball, meterstick, ruler, landing pad, collection box, white paper, carbon paper, target, tape, scale, level, C-channel clamp.

Procedures: You will be launching a steel ball into a ballistic pendulum repeatedly and making measurements in order to determine the velocity with which the steel ball leaves the projectile launcher, as in the prelab exercises. Once you are confident with your calculation of this initial velocity, predict where the steel ball will strike the floor if the ballistic pendulum is removed, as in the prelab exercises. Place a target in this position. When your instructor is available to supervise your first launch, with the ballistic pendulum now removed, launch the steel ball and see how close you come to the target.

The ballistic pendulum must capture all launches until the end of the experiment when the instructor is watching. Premature launches are not permitted.

Caution

Do not look into the barrel of the launcher. Wear safety goggles. Do not place your hand or anything else in front of the launcher. Remove the ball from the barrel of the launcher before removing the ballistic pendulum in order to prevent accidental launches. When you are finally ready to launch the ball to attempt to strike the target, be sure that space is clear – that everyone is out of the way, and knows that you are about to launch a ball so that they do not walk in the potential path of the ball.

Before launching a ball into the ballistic pendulum, ensure that the thumbscrews are tightened securely.

Caution

Do not load the ball into the barrel launcher with your finger. Use the barrel loader to do this to prevent injury.

Review your solutions to the prelab exercises in order to determine which quantities need to be measured and how to conduct the experiment. Here are a few tips:
- Clamp the projectile launcher down for increased stability.
- Check how level the projectile launcher is.
- Ensure that the ball does not return to the front edge, but rests against the spring.
- Do the entire experiment without needing to move the projectile launcher at the end of lab, in which case the ball position may change. Plan ahead.
- Use the same range setting throughout.

- Note how the length of the ballistic pendulum is defined, and that you can find the center of mass of an object approximately by balancing it on your finger.
- Zero the angle indicator, or account for any initial deviation from zero.
- Record angles to the nearest tenth of a degree.
- Check the units, which must be consistent with the units for gravitational acceleration.
- Make sure you can correctly distinguish between the different masses involved in the derivation, which you need to measure during lab.

Place a sheet of carbon paper beneath the target and make a test mark on one corner to verify that you have these stacked together properly. Place a collection box behind the target. Also, place the target and carbon paper atop a landing pad to cushion the floor against a dent, which is sturdy enough for a mark to be made on the carbon paper.

Data Check: Prior to leaving lab, the following data should be gathered:
 ✓ all quantities measured and used to compute the position of the target
 ✓ the actual position where the ball lands

Analysis: Organize clearly your complete calculation of where the ball should land.

Results: Make a table of the main quantitative results for this experiment. Include the percent error or difference.

* * * * * * * * * **What to Turn In** * * * * * * * * * *

The Report: The report for Experiment 10 should include the following sections:
 ✓ Data Table.
 ✓ Analysis.
 ✓ Tabulation of Results.
 ✓ Introduction.
 ✓ Conclusions. A minor part of your conclusions should address whether or not energy was conserved for the collision, quoting results to support your claim.

Research: Submit your completed paper. Include any necessary updates and revise your paper based on any feedback that you have already received. Exchange your paper with a peer. Review your peer's paper thoughtfully and carefully. Make some notes on your peer's draft that *suggest* conceptual or structural changes that *might* be useful. You are not grading the paper, nor are you stating what is correct or incorrect. Be suggestive. Express your constructive comments collegially. Be sure to include a few positive comments for which aspects of the paper you like the most.

Prelab Exercises for Experiment 11: Complete the following prelab exercises prior to the next lab.
 1. A fulcrum is placed beneath the 50-cm mark of a horizontal meterstick. If a 30-g mass is suspended from the 70-cm mark and a 40-g mass is suspended from the 20-cm mark, how much mass must be placed at the 60-cm mark in order to achieve rotational equilibrium?

Experiment 11: Torque

Introduction: Torque is sometimes confused with force, but torque and force are two fundamentally different physical quantities. To get a sense of why this confusion might occur, open and shut a door a few times. If the door is open and you wish to shut it, what do you need to do? You're probably thinking that you need to apply a force to shut the door. Well, you do, but this is imprecise. In fact, it is possible to apply a force in such a way that the door does not close at all. What you really need to do is apply a torque. A force is needed to cause a torque, but not all forces cause torques. It is necessary to distinguish between these two quantities.

Let's back up and return to the door. Suppose a friend asks you, "If you push the door with a force equal to your weight, do you think the door will close?" Before you answer, your friend wagers $5 that you won't be able to close the door; and you can even push as hard as you like. Ah, but your friend didn't clarify how the door was to be pushed. After you say, "You're on!" and stride over to the door with a smile on your face, your friend says, "But you must push on the hinges." Unfortunately, you won't be able to shut the door (at least, not by *pushing* on the hinges – grabbing them firmly and twisting them would not be a push, though this would be an extremely inefficient way to close a door). Your friend admits that wasn't fair, so he makes you a deal: He allows you to grab both the handles and push toward the hinges. Unfortunately, this doesn't work either.

The reasons that these forces do not close the door is that no torque is exerted on the door in these situations. When the door is pushed or pulled in a way that causes the door to rotate, a torque is applied. Another common mistake is to think that torques cause rotation, but this is just as imprecise as it is to think that forces cause motion. Recall that objects have inertia, which is a natural tendency to maintain constant velocity. A force is not required for an object to move; once moving, an object will keep moving with constant speed and direction in the absence of a net external force. Similarly, rigid bodies have a natural tendency to maintain constant angular velocity. A net force acting on an object causes the object to accelerate, while a net torque acting on a rigid body causes its angular velocity to change.

Textbook Reading: Read the sections of the textbook about torque and static equilibrium. The formula for torque and strategy for balancing torques for a rigid body in static equilibrium will be applied in this experiment. Examine the formula for torque and try to understand mathematically why the torques are zero for the examples of trying to shut the door in the introduction above.

Objective: Experimentally verify the condition for rotational equilibrium.

Apparatus: metersticks (without metal endcaps), fulcrum, meterstick clamp, torque hooks, set of slotted masses with hangers, tape, scale.

Procedures: Record the mass and location of the center of mass of the meterstick, which may be slightly off compared to the 50-cm mark. For each situation described below, experimentally determine the specified unknown quantity. Increase and decrease the unknown mass or position slightly in order to establish the uncertainties.

1. Given a load of mass m at +40 cm, what load is needed at −30 cm to achieve equilibrium?

2. Given a load of mass m at +40 cm, where must a load of $2m$ be placed in order to achieve equilibrium?
3. Given a load of mass m at +40 cm and a load of mass $2m$ at +15 cm, what load is needed at –30 cm in order to achieve equilibrium?
4. Given a load of mass m at +40 cm and a load of mass $2m$ at +15 cm, where must a load of $3m$ be placed in order to achieve equilibrium?
5. Place a load of m_1 at +40 cm and a load of m_2 at –30 cm. What must m_1 and m_2 be in order to achieve equilibrium if $m_1 + m_2 = 150$ g?
6. Place 100 g at +40 cm and move the fulcrum until equilibrium is achieved.

Data Check: Prior to leaving lab, the following data should be gathered:
  ✓ mass of meterstick, location of the center of mass of the meterstick
  ✓ mass m used and unknown load found for Step 1
  ✓ mass m used and unknown position found for Step 2
  ✓ mass m used and unknown load found for Step 3
  ✓ mass m used and unknown position found for Step 4
  ✓ mass m_1 and mass m_2 for Step 5, where $m_1 + m_2 = 150$ g
  ✓ unknown position found for Step 6

Analysis: Calculate the theoretical unknown for each experiment and compare with the experimentally measured unknown and its uncertainty.

Results: Make a table of the main quantitative results for this experiment. State results in the format $x \pm \sigma_x$. Include relevant percent errors or differences and state the relative discrepancies.

*   *   *   *   *   *   *   *   *   *   What to Turn In   *   *   *   *   *   *   *   *   *   *

The Report: The report for Experiment 11 should include the following sections:
  ✓ Data Table.
  ✓ Analysis.
  ✓ Tabulation of Results.
  ✓ Illustrations.
  ✓ Sources of Error.

Research: Get together with the peer whose paper you reviewed and exchange your ideas in a friendly, collegial manner. Be positive and suggestive. Point out what is good as well as what you feel could use improvement. Your peer will make similar comments on your paper. Do not take this as a personal attack. You may not agree with the points, so remember that these are suggestions. Appreciate the time your peer took to review your paper, and later choose which advice, if any, to take. Revise your paper.

Prelab Exercises for Experiment 12: Complete the following prelab exercises prior to the next lab.

1. Look up the formulas for the moments of inertia of the following geometric objects, assumed to have a uniform distribution of mass, about the axes described. Include a list of reliable references from which you found the formulas.
    a. solid sphere about an axis passing through its center
    b. hollow sphere with inner radius R_i and outer radius R_o about an axis passing through its center
    c. hollow sphere with infinitesimal thickness about an axis through its center
    d. solid cylinder about its axis
    e. hollow cylinder with inner radius R_i and outer radius R_o about its axis
    f. hollow cylinder with infinitesimal thickness about its axis
2. Predict which of the following factors will affect the acceleration of a *solid* rigid body that rolls without slipping down an inclined plane. In each case, explain your reasoning in terms of physics concepts and/or by describing a formula. If you deem that there is a significant effect, explain what will determine if the acceleration is greater or smaller.
    a. the mass of the rigid body
    b. the radius of the rigid body
    c. the shape of the rigid body
    d. the initial speed of the rigid body
    e. the length of the inclined plane

Experiment 12: Moment of Inertia

Introduction: Recall that mass is a measure of inertia in that the more mass an object has, the less acceleration it will receive from a net external force. What determines how easy or difficult it is to change the angular velocity of a rigid body? The answer is moment of inertia. The greater a rigid body's moment of inertia is, the less its angular velocity will change for a given net external torque.

Consider a rigid body rolling down an incline without slipping. In this case, the final velocity is less than it would be if it were sliding down a frictionless incline. In either situation, the gravitational potential energy of the object is converted to kinetic energy. For sliding without friction, there is just one kind of kinetic energy – translational kinetic energy down the incline. For rolling without slipping, there is translational kinetic energy from the motion of the center of mass down the incline in addition to rotational kinetic energy from the body's rotation. In this case, part of the work done by gravity goes into overcoming the body's inertia in order to accelerate it down the incline, while the remainder goes into overcoming the body's moment of inertia in order to increase its angular velocity (both the tangential and angular velocity must correspond in order to roll without slipping). There is no angular acceleration in the case of sliding without friction, so in this case all of the gravitational work goes into accelerating the center of mass down the incline, resulting in a larger speed at the bottom of the incline compared to rolling without slipping.

There must be friction for a body to roll without slipping, but friction should not be directly blamed for the slower final speed in the case of rolling without slipping. It is not slower because friction acts directly to slow the body down, as is the case in sliding with friction. The reason it is slower has to do with moment of inertia and rotational kinetic energy, as explained above.

To see the effect that friction has compared to the rotational effect, compare a ball rolling up an incline without slipping to a block sliding up an incline without friction. Assume that they have the same initial velocity. Thinking in terms of friction, you might expect the block to travel further – without friction there is less resistive force, right? However, the ball actually travels further up the incline, even though there is friction in the case of the ball but not the block in this example. The reason is that gravity has to overcome both the inertia and moment of inertia of the ball to slow it down, whereas with the block there is just inertia, so gravity is more effective at slowing down the block – even though in that case there is no friction.

Textbook Reading: Read about moment of inertia, rotation, rigid bodies, rotational kinetic energy, and conservation of energy for an object that is rolling without slipping. The equations for the moments of inertia of rigid bodies of various geometries and rotational kinetic energy will be used, and the strategy for conserving energy for a body that rolls without slipping will be applied.

Objective: Qualitatively analyze the effects that mass, size, and/or distribution of mass have on the acceleration of an object rolling along an incline without slipping.

Apparatus: inclined plane, rolling objects of a variety of geometries.

Procedures: Conduct experiments to allow two rolling objects to race down an inclined plane. Make qualitative comparisons based on whether or not one object beats another significantly. First, test a variety of solid spheres to deduce whether or not mass or size affect the acceleration of the sphere down the incline. Test a variety of solid cylinders for the same factors. Next, compare the effect that geometry has on the outcome, controlling any effects that other factors may have.

Data Check: Prior to leaving lab, the following data should be gathered:
- ✓ qualitative observations for solid spheres, raced two at a time
- ✓ qualitative observations for solid cylinders, raced two at a time
- ✓ qualitative observations for different geometries, raced two at a time

Analysis: Compare the results for racing rolling objects down an incline of differing mass, size, and geometry to deduce which factors have a significant effect on the outcome. Write out conservation of energy symbolically for a rigid body that starts from rest and rolls without slipping down an inclined plane. Substitute γmR^2 for moment of inertia, where γ is an overall coefficient. (However, observe that some equations for moment of inertia do not have this form. Thus, this equation will apply to just those geometries that do have this simple form.) Simplify the equation for conservation of energy, showing which factors – mass, size, or shape – affect the final speed. Let this serve as your theoretical guide for predicting which factors should have a significant effect on the outcome of the experiments.

Results: Make a table of the main qualitative results for this experiment. Do not just state which objects won which races, but also comment on how significant the difference was. As always, the results should be well-organized.

* * * * * * * * * What to Turn In * * * * * * * * *

The Report: The report for Experiment 12 should include the following sections:
- ✓ Data Table.
- ✓ Analysis.
- ✓ Tabulation of Results.
- ✓ Theory. Type a theory section for this experiment, deriving equations that will be useful to refer to in the Discussion of Results. Number the equations for easy reference.
- ✓ Discussion of Results.

Theory

In paragraph form, with equations embedded using Word's equation editor, along with at least one illustration drawn with Word's drawing tools, introduce the reader to the theory, including descriptive text and discussion of the concepts and strategy in addition to mathematical equations. Derive the equations as part of your discussion, clearly explaining the steps. Your goal is for a beginning student to be able to read your Theory section and be able to follow it. Just work symbolically in this section – do *not* plug in any numbers in *this* section (unless you are giving an instructive numerical example), since calculations are part of the Analysis section.

Prelab Exercises for Experiment 13: Complete the following prelab exercises prior to the next lab.

1. Consider a spring suspended from a rod with a mass connected to the free end.

a. What is the equation for the period?
b. Show that this equation is dimensionally correct.
c. Predict what will happen to the period of oscillation if the mass of the spring is doubled. Explain your answer.
d. Predict what will happen to the period of oscillation if the suspended mass is doubled. Explain your answer.
e. Predict what will happen to the period of oscillation if the spring constant is doubled. Explain your answer.
f. Predict what will happen to the period of oscillation if the amplitude of oscillation is doubled. Explain your answer.
g. Predict what will happen to the period of oscillation if the same equipment is transported to the moon, where gravity is reduced by a factor of 6, and the same experiment is performed there. Explain your answer.

Experiment 13: Simple Harmonic Motion

Introduction: When a spring is stretched or compressed relative to equilibrium, it exerts a restoring force in accordance with Hooke's law. The overall negative sign reflects that the spring wants to restore equilibrium: If the spring is stretched, it exerts a compressing force, whereas if it is compressed, it exerts a stretching force.

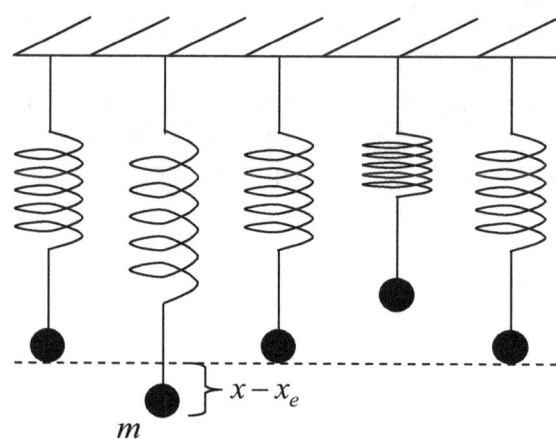

Fig. 13-1. A spring oscillates with simple harmonic motion.

Consider a spring with one end fixed and a mass connected to its free end. The spring is stretched and released from rest. The spring exerts a force toward equilibrium, compressing the spring. This causes the mass to accelerate, gaining speed and heading toward equilibrium. When the mass arrives at equilibrium, at this instantaneous moment the restoring force is zero, but equilibrium does not persist because the mass has inertia – it is moving fastest at equilibrium – so it continues past equilibrium. The spring then exerts a force to try to restore equilibrium, which decelerates the mass. The mass loses speed until coming to rest on the opposite side of equilibrium compared to where it started. It then accelerates back toward equilibrium. The result is that the mass oscillates back and forth with simple harmonic motion.

Textbook Reading: Read the sections of the textbook on Hooke's law, springs, and simple harmonic motion. Equations that will be useful include Hooke's law and the period of a spring oscillating with simple harmonic motion.

Objective: Compare the spring constant of a spiral spring obtained from two different methods.

Apparatus: spiral spring, photogate attached to stand, computer interface, set of slotted masses with hangers, meterstick, masking tape, scale, table clamp with posts, right angle clamp, horizontal rod, suspension hook, index cards, scissors.

Procedures: These procedures begin with a few experimental notes. Toward the end of the procedures is a concise statement of what data needs to be gathered.

Regarding measurements of time: Cut a slim length of index card and tape it vertically to the bottom of the base (include this as part of the suspended mass). Adjust the photogate such that the index card will properly trigger the photogate, and that only the index card triggers the photogate. Count the blinking of the red light as the index card oscillates. If too little force is applied, the oscillations will not be smooth – the motion will make an abrupt transition at the top of its motion. Try varying the mass, within maximum limits, to ensure that the oscillation is slow and smooth.

Strive for smooth vertical oscillations with a minimum of swaying and twisting. Keep a careful eye on the triggering of the photogate. Examine the periods. Keep as many consecutive periods from the start for which the period is relatively steady on average. Discard data that clearly form a downward trend. If there are statistical outliers, repeat the experiment, and consult the instructor, if necessary.

Regarding a computer interface, if used: Does the photogate with pendulum sensor expect to be triggered once or twice per oscillation? You might be measuring half or double the period. You can count the number of complete oscillations in one minute, for example, and compare with the computer software to check this. Remember to multiply by any factors of ½ or 2, if necessary.

Collect data that will enable the spring constant to be determined from the slope of a linearized graph relating to Hooke's law. Collect another set of data that will enable the spring constant to separately be determined from the slope of a linearized graph relating to the period of oscillation of the spring.

Gather data to determine whether or not the amplitude of oscillation affects the period of the spring.

Data Check: Prior to leaving lab, the following data should be gathered:
- ✓ two columns of data, with a handful of rows, relating to Hooke's law
- ✓ two columns of data, with a handful of rows, relating to the period equation
- ✓ data to determine if amplitude affects the period

Analysis: Make a plot to determine the spring constant via Hooke's law, linearizing the data if necessary. Find the uncertainty in the slope. Repeat for the period data. Compare the two spring constants.

Linearizing Data

Sometimes, data points form a linear relationship. For example, consider measurements of velocity v as a function of time t for a car rolling down an incline. In this case, the variables are theoretically related through the equation $v = v_0 + at$, where the initial velocity v_0 and acceleration a are constants. This is a linear equation since it has the form $y = mx + b$. In this case, v is playing the role of y, t serves as x, the slope represents the acceleration (since a is the coefficient t, acting as the x-coordinate), and the y-intercept is v_0. A plot of velocity as a function of time for the car rolling down the incline is therefore theoretically predicted be linear and have a slope equal to the car's acceleration.

However, sometimes data points follow a nonlinear equation. If instead position x and time t are measured for the car rolling down the incline, the plot will be curved. Position and time in this case are related through the equation $x = v_0 t + \frac{1}{2}at^2$. This equation is quadratic in time, so a plot of x as a function of t will be a parabola, not a line.

It is much easier to look at a set of data and determine whether or not it is linear than if it is a particular curve – e.g. a parabola, quartic, or exponential. Nonlinear data can be linearized in many cases through a substitution. When possible, this technique, called linearizing the data, serves as a very useful analysis tool.

Consider the case of a ball released from rest in vacuum where height h and time t are measured. These variables are related in this case by $h = \frac{1}{2}at^2$. Presently, the equation is quadratic, but it would be linear if it could be cast in the form $y = mx + b$ after an appropriate substitution, where x and y must be variables, related to the measured quantities h and t, while m and b must be constants, related to the constant a. The substitution $u = t^2$ will work in this example. Although h is quadratic in time, it is linear in u since $h = \frac{1}{2}au$. Plotting h as a function of u will result in a linear relationship with slope equal to one-half of the acceleration (the coefficient of the linearized variable u).

Results: Make a table of the main quantitative results for this experiment. State results in the format $x \pm \sigma_x$. Include the percent error or difference and state the relative discrepancy.

---

* * * * * * * * * * * **What to Turn In** *** * * * * * * * * * ***

The Report: The report for Experiment 13 should include the following sections:
- ✓ Abstract.
- ✓ Data Table.
- ✓ Analysis.
- ✓ Tabulation of Results.
- ✓ Conclusions.

Prelab Exercises for Experiment 14: Complete the following prelab exercises prior to the next lab.
1. What is the equation for the period of a simple pendulum?
    a. This equation is not exact, but an approximation. What approximation is employed in deriving this equation?
    b. Under what condition will this equation yield very good results? poor results?
    c. If the mass of the pendulum bob were increased by a factor of 2, what would happen to the period?
    d. If the length of the pendulum were increased by a factor of 2, what would happen to the period?
    e. If the pendulum were transported to the moon, where gravity is weaker by a factor of 6, what would happen to the period?

Experiment 14: Simple Pendulum

Introduction: The solution to the equation of motion for a plane pendulum is most easily solved under the following assumptions: The pendulum bob is approximately pointlike, the mass of the string is negligible compared to the mass of the bob, the string is inextensible, the pendulum swings from a well-defined point of support, the point of support is fixed, there is negligible friction between the point of support and the string, and air resistance is negligible. For those who really want to be picky, we also neglect the rotation of the earth, the variation of gravitational acceleration with altitude, etc. The list is endless; the important thing is to identify which 1-2 sources of error are most significant and quantify those.

A pendulum behaves approximately as a simple pendulum to the extent that these underlying assumptions are negligible. Even the simple pendulum, however, only approximately swings with simple harmonic motion, depending upon one particular factor (which one is left as a prelab exercise). Thus, the periodic motion of a pendulum is not quite the same as the simple harmonic motion of a spring, though in many cases the oscillatory behavior is mathematically nearly identical.

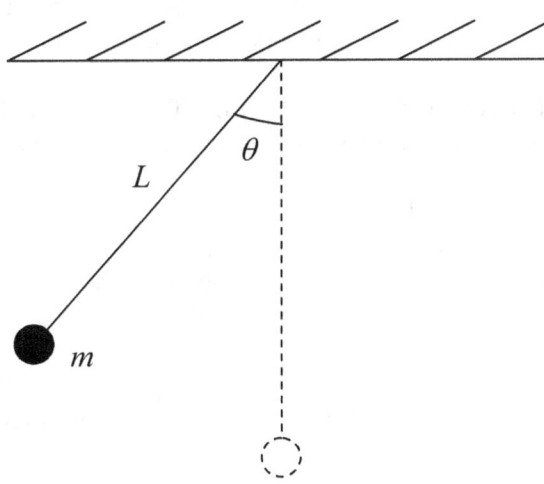

Fig. 14-1. Two different positions of a simple pendulum are shown.

Textbook Reading: Read the sections of the textbook on the simple pendulum. Pay careful attention to any approximations made in the derivations. The equation for the period of a simple pendulum, in addition to the conditions for which it applies, will be useful.

Objective: Determine which factors affect the period of a simple pendulum, and how.

Apparatus: Assorted cylindrical pendulum bobs, photogate attached to stand, computer interface, meterstick, scale, rods and clamps for pendulum support, thread, scissors, protractor, vernier calipers, masking tape.

Procedures: Perform an experiment to determine if the mass of the pendulum bob affects the period of oscillation. The length of the pendulum and amplitude of oscillation must be the same for each case. If using a computer interface, it is not necessary to enter a length since velocity will not be measured. Release the pendulum bob from rest from an easily reproducible angle.

Caution

> Be careful that the pendulum bob does not crash into the photogate head. Be ready to catch the pendulum bob as it swings back and forth if it is about to strike the photogate. Do not allow the pendulum to oscillate unattended.

Glance at the period data in table form to see if there are any statistical anomalies. There is no reason to graph the period data – this is *not* a plot to be turned in. Simply average the periods for each run. If there is a trend where the period diminishes noticeably over time, for example, then include only the beginning data points in the average.

Choose the most dense pendulum bob and now see if the length of the pendulum affects the period of oscillation. You will need a handful of data points – enough for a good plot. Reproduce the same maximum angle of oscillation for each length. Choose an angle reasonably smaller than 30°, since the equation we know is approximate. The length of the pendulum extends from the fixed point of support to the center of mass of the pendulum bob (a common mistake is to measure to the hook instead).

Now leave the length fixed and see if the amplitude of oscillation affects the period of oscillation. Note that for greater angles there is a higher risk of damage to the photogate.

Data Check: Prior to leaving lab, the following data should be gathered:
- ✓ data for period as a function of mass, with corresponding length and amplitude
- ✓ data for period as a function of length, with corresponding mass and amplitude
- ✓ data for period as a function of amplitude of oscillation, with mass and length

Analysis: Linearize the data for period as a function of length to obtain a linear graph. Predict the theoretical slope and compare to the experimental slope. Find the uncertainty in the experimental slope.

Results: Make a table of the main quantitative and qualitative results for this experiment. State results in the format $x \pm \sigma_x$. Include relevant percent errors or differences and state the relative discrepancies.

* * * * * * * * * * **What to Turn In** * * * * * * * * * * *

The Report: The report for Experiment 14 should include the following sections:
- ✓ Data Table.
- ✓ Analysis.
- ✓ Tabulation of Results.
- ✓ Discussion of Results.

Part 2: Waves, Electricity & Magnetism, and Optics

Second-Semester Sample Research Topics

Described below are 30 research projects that are extensions of topics from second-semester physics. The projects vary somewhat in challenge and complexity. Each problem includes one or more numerical goals that can be achieved through computer programming; many of these can also be solved using a spreadsheet program such as Excel. You must write the code yourself – you may *not* use part of a program written by someone else.

1. **Wave Motion**. The motion of a traveling wave satisfies the wave equation. When two waveforms overlap, the resulting wave has a form given by the principle of superposition. A train of waves traveling back and forth in a medium can result in standing waves for special relationships among the parameters. Theoretical goals: Illustrate the principle of superposition by combining the equations for two traveling waves. Numerical goals: Illustrate the principle of superposition by simulating waveforms of arbitrary shape – specified by the user as input – and visually displaying the waves as they pass each other. Extend this to a train of waves traveling back and forth in a medium.

2. **Standing Waves**. A train of sine waves propagating back and forth along a one-dimensional medium can result in standing waves, depending upon the input parameters. Theoretical goals: Derive the conditions for establishing standing waves. Numerical goals: Create a train of sine waves oscillating back and forth in a one-dimensional medium, and use the principle of superposition to form the observed wave motion. Illustrate this visually, showing what occurs in general versus the formation of standing waves. Show that the combination of input parameters that results in standing waves according to the numerical routine agrees with the theory.

3. **Complex Waveforms**. Simple sinusoidal waves of various amplitude, frequency, and phase angle can be combined to produce complex waveforms. Theoretical goals: Derive the effects of combining two sine waves that differ only in frequency, only in amplitude, or only in phase angle. Numerical goals: Write a program to combine any number of sine waves of arbitrary frequency, amplitude, and phase angle. Illustrate the resulting waveforms. Investigate the behavior of complex waveforms in a systematic way.

4. **Water Waves**. Water waves can visually illustrate waves produced by a stationary source, the Doppler effect, and shock waves. Theoretical goals: Derive the equations for the Doppler effect and shock waves specifically for water waves. Numerical goals: Write a program that produces a series of wavefronts for a source of SHM moving in water with constant velocity. The user should be able to input the frequency and magnitude and direction of the velocity of the source, and the program should graph the resulting wave patterns. Choose input parameters that demonstrate the wave pattern for a stationary source, the Doppler effect, and shock waves. Vary parameters to look for patterns, such as what happens to the half-angle of the cone of the shock wave as the speed of the source increases.

5. **Gravitational Field Mapping**. A map of gravitational field lines and equipotentials can provide great insight into a problem, especially when it cannot be solved analytically. Theoretical goals:

Consider two identical point masses lying at points $(\pm a, 0)$. Write an equation for the gravitational field at the point (x, y) and simplify it is as much as possible. Repeat for the gravitational potential at this point. An equipotential is the set of points for which the gravitational potential is a constant. Investigate the problems of setting the gravitational potential to a constant and attempting to solve for the equation of the resulting equipotential. If it were possible to obtain an equation for the equipotential, show how the gravitational field lines could easily be found. Numerical goals: Given a set of point-masses or a continuous mass distribution, write a program to find sets of equipotentials. From these equipotentials, write a program to find the gravitational field lines from the fact that the gravitational field lines should be smooth curves that are perpendicular to equipotentials that they intersect. Illustrate the resulting gravitational field maps for two point-masses, three point-masses, and other mass distributions.

6. **Electric Field Mapping**. This is similar to the Gravitational Field Mapping project, except that the charges may or may not differ in sign. Illustrate the results for the electric dipole, two positive charges, and other charge distributions.

7. **Superposition**. For a given set of point-charges, the net electric field at any given point in space is found through vector addition. Theoretical goals: Combine the equation for electric field with vector addition to derive a prescription for finding the net electric field at some specified position for a given set of point charges. In special cases, illustrate this by solving for position(s) where the net electric field is zero. Explain the difficulty in solving for such positions in general. Numerical goals: First write a program that computes the net electric field at a field point. The user will input the number of charges, the charge and location of each charge, and the position of the field point. Then write a program to find point(s) where the net electric field is zero.

8. **Gauss's Law**. Gauss's law is especially useful for computing electric fields for cases where there is a high degree of symmetry to exploit, but is otherwise very difficult to apply analytically. Theoretical goals: Illustrate how to apply Gauss's law to compute electric field for the standard geometries. Numerical goals: Write a program that numerically computes the net electric flux through any closed surface and also computes the charge enclosed by the surface. Apply the program to verify Gauss's law for the standard examples. Also apply the program to verify that Gauss's law holds for an arbitrary Gaussian surface – i.e. the user can input the surface, and the program will verify that Gauss's law holds for it. For example, the program might use a cube for a Gaussian surface for a spherical charge distribution (this is not a practical choice for applying Gauss's law by hand, but is important in showing that the Gaussian surface can be arbitrary, in principle).

9. **Charge Distributions**. Sprinkle some charges on a conductor and they very quickly reposition themselves in order to achieve electrostatic equilibrium. Theoretical goals: For simple cases, such as a circular arc or thin spherical shell, choose the location of one or two charges, as needed, and then compute the location of the remaining charges needed to achieve electrostatic equilibrium. Numerical goals: Given a 1D, 2D, or 3D conductor (specified by a function) and a specified number of charges, write a program that chooses the position of one or two charges, as needed, and then computes the location of the remaining charges needed to achieve electrostatic equilibrium. Verify that this agrees with the theory, then also apply it for other shapes. The resulting charge distributions should satisfy the qualitative expectations for conductors in electrostatic equilibrium.

10. **Electric Current**. Electric current can be modeled approximately by a classical model for electrical conduction (the textbook presents such a microscopic picture of conduction). Theoretical goals: Work out the theoretical details of this classical model for electrical conduction. Numerical goals: Apply this model to create a visual simulation of electric current. Also, perform various numerical calculations and check that they agree with the theoretical equations in appropriate limits. Systematically investigate the effects of varying the input parameters.

11. **Kirchhoff's Rules**. A general circuit with batteries and resistors can be solved via Kirchhoff's rules. Theoretical goals: Explain Kirchhoff's rules and work through a couple of examples algebraically to illustrate their application. Numerical goals: Write a program that allows the user to input a circuit with a specifiable number of batteries and resistors. The program must determine how many independent equations and unknowns are needed and how to set up the system of equations such that the equations are actually independent; it should also check that the system can be solved, and produce an error statement and end the program otherwise. The program will then solve the system of equations to find the unknown currents, including direction. Apply the program to some textbook examples to ensure that it works correctly.

12. **Magnetic Field Mapping**. This is similar to the Electric Field Mapping project, except that it is applied to bar or horseshoe magnets. Also, explore the interesting region inside the magnet, where the magnetic field lines can change direction abruptly and behave much differently than they do outside the magnet; this is a striking area where the magnetic field lines of a bar magnet differ from the electric field lines of a dipole, and nicely illustrates Gauss's law in magnetism.

13. **Moving Charges**. If the net force exerted on a moving charge arises solely due to a uniform magnetic field, then the path of the charge will be a straight line, a circle, or a helix depending upon its initial velocity and the direction of the magnetic field. Theoretical goals: Prove that the possible paths of motion are a straight line, a circle, or a helix. Derive conditions that determine which path results. Numerical goals: For input consisting of the magnitude and direction of a uniform magnetic field and the initial velocity of a moving charge as well as the value of the electric charge, write a program that graphs the resulting motion of the charge. Extend the program to accommodate additional forces, such as the weight of the charge or a uniform electric field.

14. **Ampère's Law**. Ampère's law is especially useful for computing magnetic fields for cases where there is a high degree of symmetry to exploit, but is otherwise very difficult to apply analytically. Theoretical goals: Illustrate how to apply Ampère's law to compute magnetic field for the standard geometries. Numerical goals: Write a program that numerically computes the sum of $B_i \Delta s_i \cos\theta_i$ over the elements of any closed path and also computes the current enclosed by the path. Apply the program to verify Ampère's law for the standard examples. Also apply the program to verify that Ampère's law holds for an arbitrary Ampèrian loop – i.e. the user can input the curve for the closed path (e.g. an ellipse), and the program will verify that Ampère's law holds for it.

15. **Electric and Magnetic Flux**. Electric flux is involved in Gauss's law, and magnetic flux is involved in Faraday's law. For simple problems, electric or magnetic flux can be calculated exactly, but numerical techniques are required to solve the most general problems. Theoretical goals: Derive equations for electric or magnetic flux through various surfaces (at least some of which must be open) for an electric or magnetic field as a specified function of position. Numerical goals: Write

a program to compute the electric or magnetic flux through a specified surface for an electric or magnetic field as a specified function of position. Show that the results agree with the theoretical derivations.

16. **Faraday's Law**. It is easiest to compute the current induced in a loop of wire for which the magnetic field is uniform throughout the loop, but numerical techniques are required for the most general problems. Theoretical goals: Derive equations for the current induced in a loop of wire of specified shape for a magnetic field as a specified function of position. Numerical goals: Write a program to compute the current induced in a loop of wire of specified shape for a magnetic field as a specified function of position. Show that the results agree with the theoretical derivations.

17. **Rainbows**. It takes some care to illustrate the formation of a rainbow by hand; it can be illustrated more precisely by the computer. Theoretical goals: Given the equation for the line of a ray of incident white light with negative slope and the equation of a circle to represent a spherical raindrop for which the light will strike the upper left corner, along with the index of refraction of water for red and violet wavelengths, determine the equations of the refracted, reflected, and once again refracted rays for both red and violet. Numerical goals: Accept as input the slope and y-intercept of an incident ray of white light, and the radius and coordinates of the center of a circle to represent a spherical raindrop. The index of refraction of water for the wavelengths of the different visible colors should be incorporated into the program, rather than user input. The program should compute the equations of the refracted ray, one reflected ray (for the primary rainbow) or two reflected rays (for the secondary rainbow), and then the final reflected ray for the different colors of visible light. Graph this data to show the formation of the primary and secondary rainbows. Create a second graph (not a drawing by hand) that shows the bigger picture: This should include parallel rays of white light corresponding to different colors of visible light, which strike raindrops at appropriate altitudes, followed by the output rays that a particular observer actually sees when viewing the primary rainbow.

18. **Fermat's Principle**. According to Fermat's principle, light takes the path of least time. The law of reflection and Snell's law follow from Fermat's principle. Theoretical goals: Apply Fermat's principle to justify the law of reflection and Snell's law for a planar interface between two media by computing the total trip time for various routes. Numerical goals: Given the coordinates of initial position in one medium and final position in another medium, the indices of refraction of the two media, and an equation for the interface, write a program to compute the total time that it would take light to travel from the initial position to the final position along two straight line segments starting from all possible directions from the initial position to determine which route results in the minimum time. Verify that the results are consistent with Snell's law.

19. **Huygens' Principle**. Huygens' principle offers a means of constructing a new wavefront from an old wavefront: Each point on the old wavefront serves as a point source for a spherical wavelet, and the new wavefront, at some later time, is the surface tangent to those wavelets. Theoretical goals: Apply Huygens' principle to derive the law of reflection and Snell's law. Numerical goals: First, write a program that takes as input the equation for a wavefront and time interval, employs Huygens' principle to create spherical wavelets, and after the specified time interval determines the surface of the new wavefront that is tangent to these wavelets. Apply this to a plane wave and spherical wave to test the program. Then apply Huygens' principle for a plane wave incident upon

an interface, and graph the resulting reflection and refraction. Finally, apply Huygens' principle for a plane wave incident upon a single or double narrow slit, and plot the resulting interference pattern.

20. **Curved Mirrors**. The standard textbooks treat the spherical mirror because it is simplest, but there is actually some aberration. The mathematics for the parabolic mirror is somewhat more involved, but the parabolic mirror eliminates aberration. Theoretical goals: Show that the spherical mirror formulas in the textbook are only approximate, and prove that there is actually aberration. Investigate the limits in which the aberration is better or worse. Show that there is no aberration for the parabolic mirror. Numerical goals: First work with parallel rays of light incident upon the surface of a mirror, for which the equation is specified as a particular parabola, ellipse, circle, or hyperbola by the user, and numerically determine the reflected rays. Graph the incident and reflected rays. This should demonstrate that there is aberration except for the parabolic mirror. Next, specify an object distance and height rather than working with incident parallel rays. Create rays that radiate outward from the top of the object and graph their reflection by a specified mirror to show the image formation.

21. **Lens Systems**. One or more lenses working together form a lens system. The image of one lens serves as the object for the next lens. Theoretical goals: For two lenses, derive an equation for the final location and height of the final image in terms of the focal lengths of the lenses, the initial object position and height, and the distance between the lenses. Also, work out the overall magnification and prove that it is the product of the individual magnifications. Numerical goals: Write a program that computes the location and height of the final image for a user-specified number of lenses, lens positions, object position, and object height. The program should also determine the overall magnification of the system and the orientation and character of the final image. Verify that the program agrees with some odd-numbered textbook problems.

22. **Thick Lenses**. The refraction through the lens does have an effect, which is significant especially for thick lenses. Theoretical goals: Given the equation of a ray incident upon a thick lens and the equations of the two spherical sides of the lens, determine the equations of the ray that refracts through the lens and the ray that emerges from the lens. Numerical goals: First work with parallel rays of light incident upon a thick lens, for which the equations of the circular sides are specified by the user, and numerically determine the rays that refract through the graph and the rays that emerge on the opposite side. Plot the ray diagram. Compare to the ray diagram for a comparable thin lens. Next, specify an object distance and height rather than working with incident parallel rays. Create rays that radiate outward from the top of the object and again determine the rays that refract through the graph and the rays that emerge on the opposite side. Plot the image formation. Compare to the image formation for a comparable thin lens.

23. **Aberrations**. Spherical mirrors have aberrations and lenses have both spherical and chromatic aberrations. Theoretical goals: For incident parallel rays of light, show that there are spherical aberrations in both mirrors and lenses as well as chromatic aberrations in lenses. Numerical goals: First work with parallel rays of light and a choice of spherical mirror or lens with two spherical sides and compute the output rays. Plot the ray diagrams to show the spherical aberrations. Next, compute the output rays for incident parallel rays of white light and a lens with two spherical sides to show the chromatic aberrations. Finally, specify an object distance and height rather than

working with incident parallel rays. Create rays that radiate outward from the top of the object. Repeat the ray diagrams for spherical and chromatic aberrations.

24. **Fiber Optics**. The underlying principle behind fiber optics is total internal reflection. Theoretical goals: Given the equation of an incident ray and the equation of an interface between two media, assuming that the first medium has the higher index of refraction, derive an equation for that determines whether or not there will be total internal reflection. Numerical goals: Given the equation for an incident ray in glass, the index of refraction of the glass, and the equations for the boundaries of the glass, numerically map out the path that the beam of light follows as it travels through the glass, and at any point where the beam strikes the surface of the glass, determine whether or not the beam partially refracts. Draw the resulting ray diagram, including any partial refractions.

25. **First Emergence**. When a ray of light in air enters a piece of glass, it may reflect multiple times with total internal reflection before any part of the light finally emerges from the glass. When it finally does emerge, this is called the ray of first emergence. Theoretical goals: For an equilateral triangle and a square, separately, given the equation of an input ray, work through an example where the ray does not emerge at the first chance, deriving an equation for the ray of first emergence. Numerical goals: Given the equation of an input ray, the vertices of a convex polygon, and the index of refraction of the glass, numerically solve for the equation of the ray of first emergence. Plot the ray diagram up to and including the first emergence.

26. **Multi-Slit Interference**. The interference pattern for three or more slits features some interesting fringes compared to the double-slit pattern. Theoretical goals: Work out the theory for triple-slit interference analogous to the theory for double-slit interference, including the conditions for constructive and destructive interference and applying the phasor addition of waves to obtain the intensity distribution. Explain the extra in-between fringes that appear in the multi-slit patterns. Numerical goals: Given the slit width, slit spacing, number of slits, wavelength of the light, and distance between the slits and screen, numerically solve for the intensity pattern that appears on the screen. Graph the intensity as a function of position on the screen. Explore the effects of changing the input parameters.

27. **Interferometry**. An interference pattern is formed when a beam of light divides into two paths and recombines at a later point. Such a device is called an interferometer and has many practical uses. Theoretical goals: Work out the theory of the interference patterns formed by the Michelson interferometer. Numerical goals: Given input for the Michelson interferometer, numerically calculate the interference effects to plot the resulting interference pattern – which consists of bright and dark circular rings.

28. **Diffraction Grating**. This is similar to the Multi-Slit Interference project, except that it is instead applied to the diffraction grating.

29. **Angle of Deviation**. A prism is very useful for dispersing light – i.e. separating a beam of light into its constituent wavelengths. Theoretical goals: For the case that a beam of monochromatic light is incident upon a prism in such a way that the ray emerges symmetrically, derive an equation for the angle of deviation in terms of the prism angle (only one of the three angles will be relevant) and the index of refraction of the glass – no other variables may appear in your final expression. Do not assume a known shape for the triangle – e.g. do not assume to know the other two angles that are not needed, and do not assume that it is isosceles either. Numerical goals: Given the equation for the incident ray, the one needed prism angle, and the index of refraction of the glass, numerically determine the angle of deviation (do not assume it to pass through symmetrically like the special case treated in the theory). Plot the ray diagram. Use your program to show that the angle of deviation is actually minimum when the ray passes through symmetrically. Verify that the theory and program agree in this case.

30. **Ray Tracing**. The ray tracing for a system of lenses can be formulated with matrix methods. Theoretical goals: Work out the theory of these matrix methods for drawing ray diagrams for a system of lenses. Work out a couple of examples. Numerical goals: Write a program to apply the matrix method formalism. For given input systems of lenses, plot the resulting ray diagrams.

Experiment 15: Standing Waves

Introduction: When a string is connected to a mechanical oscillator, the simple harmonic motion of the oscillator sends a continuous train of sine waves along the string. For almost all values of the tension in the string and frequency of the oscillator, nothing special will occur – there will just be a little wiggling of the string, with an amplitude no greater than that of the oscillator (about a millimeter). However, if the tension and frequency match just right, the train of sine waves can reinforce one another upon reflection, creating a large amplitude (a few centimeters) standing wave.

Textbook Reading: Read the sections of the textbook on traveling waves, wave superposition, waves on a string, and standing waves. Two different equations for wave speed will be useful, in addition to the problem-solving strategy for standing wave problems.

Objectives: Produce standing waves on a string, and compare the speed of the waves using two different formulas.

Apparatus: mechanical oscillator, electrical wires with banana leads, function generator, set of slotted masses with mass hanger, pulley with table clamp, table clamp with posts, rods, BNC to banana plugs connector, alligator clip, meterstick, thread or string, scissors, scale. Alternatively, standing waves may be set up in an air column using a tuning fork and rubber mallet.

Procedures: The relative error in measuring the mass of the thread or string can be reduced if the lab collaborates by weighing all of their thread or string together. Each group will need about 1.5 m of thread or string. On some digital scales, a more precise measurement of mass can be obtained in ounces, at the minor cost of needing to convert afterward.

Record the total length of the string. Set up the experiment to send traveling waves along the string in such a way that the tension can be determined. Record the active length of the string. These two different lengths will both be useful.

Connect the function generator to the mechanical oscillator, leaving the power off until the instructor approves your circuit. Start with the amplitude at zero so that the current will not be high when the power is turned on.

Decide whether to vary the frequency or the tension, with the intention of leaving the other fixed (though some values of either frequency or tension may not work out well). Find the fundamental resonance and the first few overtones. Once a standing wave is found, increase and decrease the varied parameter (either frequency or tension) a little in order to determine the best value and its uncertainty. The equations can serve as a guide as to whether frequency (or tension) needs to be increased or decreased in order to find the remaining standing waves.

For each standing wave, sketch the wave, record the frequency and a physical quantity related to the tension, and measure the amplitude.

Data Check: Prior to leaving lab, the following data should be gathered:
- ✓ total and active lengths of the thread or string, mass of the thread or string
- ✓ sketches, frequency, data related to tension, and amplitude for a handful of standing waves

Analysis: Calculate the mass per unit length for the thread. Calculate the experimental and theoretical wave speed corresponding to each suspended mass for the thread. The theoretical wave speed depends on the wavelength, while the experimental wave speed depends on the suspended mass.

Results: Make a table of the main quantitative results for this experiment. State results in the format $x \pm \sigma_x$. Include relevant percent errors or differences and state the relative discrepancies.

* * * * * * * * * * **What to Turn In** * * * * * * * * * * *

The Report: The report for Experiment 15 should include the following sections:
- ✓ Abstract.
- ✓ Data Table.
- ✓ Analysis.
- ✓ Tabulation of Results.
- ✓ Sources of Error.

Research: Choose a research topic. Find one journal article or textbook (other than the required course textbook) at a suitable reading level that relates to your research project. Xerox no more than five pages that are relevant for your research topic. Attach a cover sheet to the xerox pages to submit as the first component of your research project (you may wish to keep a second copy for yourself). The cover sheet should include the subject of your research project, a brief description of your project, and a reference to the journal article or textbook in an appropriate bibliographic style.

Prelab Exercises for Experiment 16: Complete the following prelab exercises prior to the next lab.
1. Design an experiment to investigate the interactions between electrically charged objects. Include:
   a. a precise statement of the primary experimental objective ("the interactions between electrically charged objects" being rather vague); a variety of specific goals may fall into this class
   b. a list of equipment that is likely to be available; some common household materials, such as aluminum foil, can be brought to the lab, if needed
   c. a set of procedures that thoroughly describe the details of how the experiment will be conducted, including which quantities will be measured and how the measurements will be made; this may include qualitative observations
   d. a description of how the needed analysis will be performed

Experiment 16: Electric Charge

Introduction: Comparing Coulomb's law for electric charges to Newton's law of universal gravitation, it might seem that electricity should be very similar to gravity. These force laws do have very similar form in that they are both inverse-square laws, and each force is proportional to the source – electric charge being the source of an electric field, mass the source of gravity. However, there are a few fundamental differences for which electrical phenomena and gravitational phenomena are vastly different.

For one, all masses attract one another, whereas electric charges sometimes repel and sometimes attract. Related to this point, mass is never negative, but electric charge can be positive or negative. An object can also have zero electric charge, in which case it does not feel attracted to nor repelled by charged objects. There are objects with zero rest-mass, like photons (particles of light), but such particles are never found at rest (photons travel lightspeed); in fact, such particles have relativistic mass (a consequence of Einstein's theory of relativity), and actually bend very slightly due to strong gravitational fields (well, the effect can be strong near the event horizon of a black hole, but otherwise the effect is generally slight).

All masses are positive, and any two of these positive masses attract. Charges of the same sign, on the other hand, repel – only opposite charges attract. The universe would be drastically different if masses repelled, and similarly if like charges attracted and opposites repelled. This difference in attraction and repulsion is a fundamental difference between the two interactions.

Mass actually plays two roles: On the one hand, mass is a measure of inertia, but it is also the source of a gravitational field. Electric charge, however, is only the source of an electric field, not a measure of inertia (the net external force will never equal charge times acceleration).

Also, the proportionality constant, expressed in SI units, is a very small power of 10 in the case of gravity ($\sim 10^{-10}$), but a very large power of 10 for electricity ($\sim 10^{10}$). In addition, the basic unit of charge is the charge of a proton, for which the power of 10 is $\sim 10^{-19}$, while the mass of a proton has a power of 10 of $\sim 10^{-27}$. Putting this all together, the electrical repulsion between two protons is roughly 10^{36} times greater than their gravitational attraction. That's a trillion times a trillion times a trillion. As a result, it takes one astronomical mass to make an appreciable (on the order of a Newton) gravitational field, so the gravitational interaction is generally negligible in comparison to other forces when one of the masses is not astronomical. The electrical interactions of charged particles, however, is very significant even at the microscopic level.

Textbook Reading: Read about Coulomb's law, electric charges, conductors in electrostatic equilibrium, the electroscope, and Faraday's ice pail in the textbook.

Objective: Design and conduct an experiment to investigate the interactions between electric charges.

Apparatus: equipment and supplies needed to study the interactions between electric charges.

Procedures: Once you obtain the approval of the instructor, carry out the experiment you designed as a prelab exercise. For any experiment that involves an electric circuit, leave the power off and seek the approval of your instructor before turning power on.

Data Check: Prior to leaving lab, the following data should be gathered:
- ✓ Make a list of what data you need in order to meet the objectives of your experiment. Record qualitative observations as well as quantitative data.

Analysis: Perform any calculations with quantitative data that are needed in your analysis of your experiments.

Results: Make a table of the main quantitative (or qualitative) results for this experiment. State any quantitative results in the format $x \pm \sigma_x$.

* * * * * * * * * * * **What to Turn In** * * * * * * * * * * *

The Report: The report for Experiment 16 should include the following sections:
- ✓ Data Table.
- ✓ Analysis.
- ✓ Tabulation of Results.
- ✓ Experimental.
- ✓ Discussion of Results.

Research: Type the introduction to your research project.

Prelab Exercises for Experiment 17: Complete the following prelab exercises prior to the next lab.

1. A 50-μC charge is located at $(20 \text{ cm}, 0)$ and a -50-μC charge is located at $(-20 \text{ cm}, 0)$. Find the net electric field and potential at:
    a. $(0,0)$
    b. $(10 \text{ cm}, 0)$
    c. $(-10 \text{ cm}, 0)$
    d. $(0, 10 \text{ cm})$

Experiment 17: Electric Field and Potential

Introduction: Equipotentials can be mapped out experimentally by connecting a DC power supply to two electrodes painted onto carbon-impregnated paper. Leaving the ground wire of a voltmeter permanently connected to the negative electrode, the free probe can then be utilized to locate a set of a points where electric potential is constant. Once the equipotentials have been mapped, smooth curves can be drawn for electric field lines from the knowledge that electric field lines are perpendicular to equipotentials when they intersect, and also from the fact that electric field lines do not intersect each other.

Although the electric field map primarily serves as a qualitative device, it also provides quantitative information. In principle, electric field can be computed at any point on the diagram from the superposition of the electric fields due to each electrode. This is not useful in practice for an electric field map, however, since one or more electrodes are often not pointlike, and can even have irregular shapes. However, electric field can also be computed based on its relationship with electric potential.

Where the electric field is approximately radial, its magnitude is approximately

$$E \approx \lim_{\Delta r \to 0} \frac{\Delta V}{\Delta r}$$

For small Δr,

$$E \approx \frac{\Delta V}{\Delta r}$$

where Δr is the distance between consecutive equipotentials and ΔV is the difference in electric potential (a.k.a. potential difference, or voltage) between the consecutive equipotentials. To find the electric field at an arbitrary point X on the diagram, draw Δr as a straight line through X that connects the consecutive equipotentials such that the direction of Δr is the average direction of the electric field line passing through between the equipotentials, and use the formula above.

Textbook Reading: Read the sections of the textbook on electric field, electric field lines, superposition of electric fields, electric potential, the electric dipole, and batteries and DC power supplies. A basic equation to know is the one for the electric force exerted on a test charge in an electric field.

Objective: Map out the equipotentials and electric field lines for a system of two oppositely charged conductors.

Apparatus: DC power supply, corkboard, carbon-impregnated paper, carbon copy paper, blank white paper, conductive ink, conductive thumbtacks, alligator clips, multimeter with free probe, electrical wires, ruler.

Procedures: At the beginning of lab (before the instructor discusses the lab), use the conductive ink to draw your electrode configuration (an example is shown in Fig. 17-1) on the carbon-impregnated paper. Follow the instructions on the conductive ink container. A point-like electrode should be about the size of a dime (no larger), and the linear electrode should be half, or a little more, as thick. There should be plenty of room on the sides of the electrodes to explore the surrounding area, as well as enough room between them to explore the interior region. There may be problems if the ink is too thin or if the ink is not dry prior to lab. The conductive ink should be smooth. You might fan it to speed up the drying process.

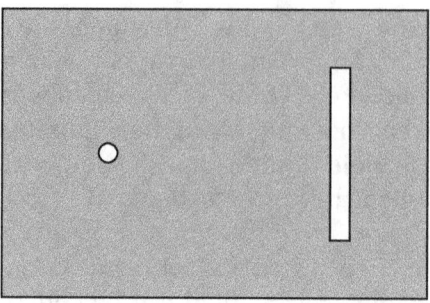

Fig. 17.-1. A sample electrode configuration is shown.

Place two sheets of blank white paper and two sheets of carbon copy paper beneath the carbon-impregnated paper, with the electrode configuration face up. Fasten these papers to the cork board via thumbtacks. Place a conductive thumbtack in each of the electrodes. Puncture the electrode in a region where the conductive paint is thick and dry. Be careful not to allow the thumbtack to wiggle relative to the electrode, or the quality of the connection may be reduced and alter readings during the experiment. If this occurs, first turn the power supply off and then move the thumbtack in a different location. Press the free probe firmly on one corner of the carbon-impregnated paper to make a test mark, then inspect each sheet of blank white paper to ensure that the mark shows.

Set up the circuit according to the schematic diagram in Fig. 17-2. Leave the power supply off until the instructor approves the circuit. Set the voltage and current to zero before turning the power supply on. Once the power supply is turned on, gradually adjust the voltage and current knobs until the power supply voltage is 10 V.

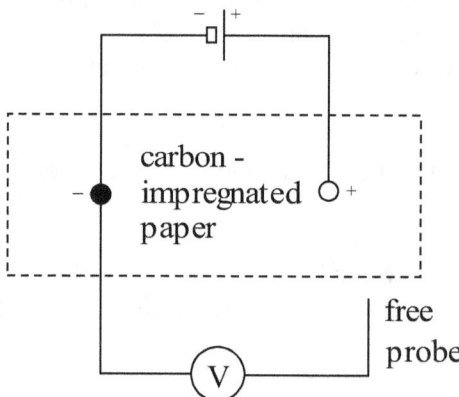

Fig. 17-2. A schematic diagram is drawn for the electric field mapping experiment.

Note

Get your instructor's attention and wait patiently if you encounter any problems during lab. A common symptom of a poor connection is for the voltage readings to suddenly change midway through lab. If the change is significant, the problem may be solved by turning the power supply off and repositioning the thumbtacks (keeping in mind that too many holes in a small area will introduce a new problem). If this does not work, please wait patiently for your instructor's assistance.

Trace the outline of each electrode with the free probe. Gently tap the free probe on the carbon-impregnated paper to achieve a voltmeter reading without making a mark on the sheets of blank white paper. Sample different regions to see how the voltage is distributed. The voltage should change by several volts going from near one electrode to near the other (*near* meaning without touching). If this is not the case, see the note above. Choose a voltage (e.g. 5 V), and move the free probe until the voltmeter reads this value. Press firmly at this point to make a mark on the sheets of blank white paper. Repeat for several points of this voltage until you have traced out this equipotential. Repeat for a handful of other voltages until you have equipotentials spread out throughout the paper. Neatness is important, as is achieving a good spread of curves. In your laboratory notebook, make a sketch of the curves you are tracing out as your experiment progresses to label the voltages so that you can add these values to your finished maps later.

Have your instructor draw an X somewhere on your diagram before you leave lab.

Data Check: Prior to leaving lab, the following data should be gathered:
- ✓ map of several equipotentials and outlines of the electrodes
- ✓ notebook sketch of equipotentials with labels for voltages and signs of electrodes
- ✓ the instructor's X on your map

Analysis: Draw smooth curves through your sets of equipotential points (do *not* connect-the-dots). Label the voltages on your diagram. Label the positive and negative electrode. Draw several smooth electric field lines throughout your map that run from the positive electrode to the negative electrode. Add arrows to indicate direction. Remember that electric field lines are perpendicular to equipotentials – so ensure that all intersections make a 90° angle.

Draw Δr through the instructor's X. Calculate the electric field at point X and show your work on your map near the X. Determine the uncertainty in the electric field at point X.

A good electric field map will contain:
- electrode outlines
- smooth curves representing equipotentials spread throughout your paper (averaged between dots that represent original data points)
- smooth electric field lines spread throughout your paper
- positive and negative signs on electrodes
- voltage labels for equipotentials and electrodes
- your instructor's X and a line and label for Δr

Results: Make a table of the main quantitative results for this experiment. State results in the format $x \pm \sigma_x$.

* * * * * * * * * *   What to Turn In   * * * * * * * * * *

The Report: The report for Experiment 17 should include the following sections:
- ✓ Data Table.
- ✓ Analysis.
- ✓ Tabulation of Results.
- ✓ Introduction.

Research: Develop the formalism for your research project.

Prelab Exercises for Experiment 18: Complete the following prelab exercises prior to the next lab.
1. Find the equivalent resistance for each of the following circuits:
    a. A 33-Ω resistor and 100-Ω resistor are connected in series.
    b. A 33-Ω resistor and 100-Ω resistor are connected in parallel.
    c. A 33-Ω resistor and 100-Ω resistor are connected in parallel, and this parallel combination is connected in series with a 10-Ω resistor.

Experiment 18: Series and Parallel Resistance

Introduction: The formulas for series and parallel equivalent resistances follow from Kirchhoff's rules. Kirchhoff's junction rule expresses conservation of charge; since the amount of charge entering a junction at any moment must equal the amount of charge leaving the junction at that moment, it follows that the sum of the currents entering the junction equals the sum of the currents leaving the junction. Although the fundamental conserved quantity is electric charge, it is more convenient to work with current (the instantaneous rate of flow of charge) for circuits with resistances, since Ohm's law involves current.

Kirchhoff's loop rule follows from conservation of energy. Energy is the fundamentally conserved quantity, but quantitatively the rule is expressed in terms of electric potential difference; again, this is out of convenience, and possible since potential difference is related to electric potential energy. In particular, energy is the ability to do work, and electric potential is defined based on how much work would be done moving a test charge between two points in a circuit. Specifically, Kirchhoff's loop rule states that the sum of the potential differences around a closed loop in a circuit must be zero.

The equation for parallel equivalent resistance reflects that resistors in parallel have the same potential difference, as required by Kirchhoff's loop rule, and that the current entering the combination equals the sum of the currents through the various resistors, according to Kirchhoff's junction rule. For series equivalent resistance, there is no junction, so there is only one current to speak of, and the sum of the potential differences across the resistors equals the potential difference across the combination in order to satisfy the loop rule.

Potential difference and current can be measured in lab using multimeters. A multimeter used as a voltmeter measures potential difference. A voltmeter has a very large internal resistance, so as to only draw a negligible amount of current, and connects in parallel between two points in a circuit. Since the voltmeter has a huge internal resistance, if it is instead connected in series with any circuit element, it has the undesired effect of removing that element from the circuit.

When used as an ammeter, a multimeter measures current. Ammeters have very tiny internal resistances, so it is very easy to inadvertently blow the fuse. Ammeters must be wired in series as an integral part of the circuit. When an ammeter is instead connected in parallel with a circuit element, since a greater percentage of the current takes the path of less resistance and since the ammeter has almost no resistance, the ammeter will draw almost all of the current – thereby blowing the fuse.

A schematic diagram is a convenient method of drawing an electric circuit. Consider the schematic diagram illustrated in Fig. 20-1. Table 20-1 indicates what these different schematic symbols represent. The dots in Fig. 20-1 indicate where explicit connections are made at a junction; these connection points are sometimes omitted when the junction should be clear from the context. The positive and negative terminals are often not labeled, since the longer line at one side of the schematic symbol represents the positive terminal (the small rectangle is often drawn as a short line).

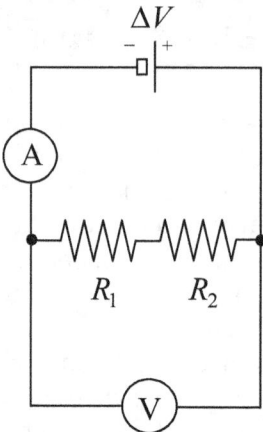

Fig. 20-1. A schematic diagram for two resistors connected in series with a battery.

−□|+	⋁⋁⋁	(A)	(V)
battery or DC power supply	resistor	ammeter	voltmeter

Table 20-1. Some common schematic symbols are defined.

Textbook Reading: Read the sections of the textbook on resistance, potential difference, current, Ohm's law, Kirchhoff's rules, and series and parallel resistance. Ohm's law, Kirchhoff's rules, the equations for series and parallel resistance, and the strategy for simplifying a circuit that consists of series and parallel resistors are relevant for this experiment.

Objective: Compare the theory for simplifying series and parallel combinations of resistors with experiments with DC circuits.

Apparatus: DC power supply, circuit board, three resistors, multimeters (or voltmeter and milliammeter, or voltage sensor and current sensor), electrical wires.

Procedures: Connect the DC power supply across a single resistor similar to the schematic diagram of Fig. 20-1. Do not use the smallest of the three resistors until the final circuit, which involves three resistors. Wait for your instructor to approve the circuit before turning power on.

Caution

The current and voltage knobs should be turned down before powering the circuit with the DC power supply. Otherwise, when power is turned on you might be inadvertently providing more current than the equipment can handle. If the meter turns off during lab, immediately turn the power supply off – before resetting the meter. Do not rotate the selection knob on the meter while the meter is connected.

Caution

Be very careful not to exceed the maximum current that the ammeter can handle. It is very easy to blow the fuse. Sometimes, it may seem like the knob is not working, and then suddenly when it does work the current goes from really tiny to way too much. Even when used properly, the current can jump too high with a seemingly slight turn of the knob. Monitor the digital readout as you adjust the power supply voltage – do not exceed the ammeter's threshold.

Caution

Any shorts in the circuit will blow the fuse in the ammeter. For example, if you try to connect a wire across a resistor, this is a short circuit: Since more current takes the path of least resistance, virtually all of the current will bypass the resistor, choosing the wire instead, resulting in much more current than the instruments can handle.

Begin with zero current and voltage. Once your circuit is approved, gradually turn the current and voltage up, being careful *not* to approach the maximum voltage written on the RLC board. Check the meter sensitivities to ensure that you stay within their tolerances. Do *not* exceed the maximum current or potential difference specified by the instructor or RLC circuit board (or other equipment used). Unfortunately, it is very easy to turn the knob just a little and quickly exceed the maximum current, so you must exercise caution. In particular, if it seems like nothing is happening, do *not* lose patience and crank the current or voltage up.

Gather pairs of current/voltage readings that will yield a good plot. Remember to record uncertainties.

Caution

Turn the voltage and current to zero and turn the power supply off before altering any connections. Do this also when the power supply is not in use. Failure to do this can easily result in a blown fuse or damage to the power supply. Simply moving a wire from one position to another can easily short the circuit, causing damage to the equipment. Do not turn the power supply on again until the instructor approves changes to your circuit.

After turning the voltage and current to zero and turning the power supply off, connect the resistors in series, but await your instructor's approval for this new circuit before turning the power supply on. You just need to make a subtle change to the existing circuit. Repeat your measurements for this series combination. Sketch the experimental setup.

Caution

Seek your instructor's approval before powering a modified circuit. If your series circuit is approved, but you reconnect it in parallel without seeking your instructor's approval, for example, you run a serious risk of blowing a fuse in the ammeter.

After turning the voltage and current to zero and turning the power supply off, connect the resistors in parallel. You just need to make a subtle change to the existing circuit. Repeat your measurements for this parallel combination. Sketch the experimental setup.

After turning the voltage and current to zero and turning the power supply off, connect a third resistor in series with the parallel combination. If using the RLC circuit board, connect the 33-Ω resistor in parallel with 100-Ω resistor, and connect this parallel combination in series with the 10-Ω resistor. Obtain your instructor's approval before turning the power supply on. Repeat your current/voltage readings.

Data Check: Prior to leaving lab, the following data should be gathered:
- ✓ sketches of the experimental setups (not schematic diagrams)
- ✓ current and potential difference data that will yield a good plot for a single resistor
- ✓ current and potential difference data that will yield a good plot for the series combination
- ✓ current and potential difference data that will yield a good plot for the parallel combination
- ✓ current and potential difference data that will yield a good plot for the parallel/series circuit

Analysis: For each resistor or resistor combination, determine the equivalent resistance in terms of the slope of a linearized graph. Also, compute the uncertainties.

Results: Make a table of the main quantitative results for this experiment. State results in the format $x \pm \sigma_x$. Include relevant percent errors or differences and state the relative discrepancies.

***   *   *   *   *   *   *   *   *   *   **What to Turn In**   *   *   *   *   *   *   *   *   *   *   ***

The Report: The report for Experiment 18 should include the following sections:
- ✓ Data Table.
- ✓ Analysis.
- ✓ Tabulation of Results.
- ✓ Sources of Error.
- ✓ Conclusions.

Research: Draw and label professional-looking illustrations relevant for your research project.

Prelab Exercises for Experiment 19: Complete the following prelab exercises prior to the next lab.
1. Copy the schematic diagram of Experiment 19. Draw and label currents in each branch of the circuit.
2. Apply Kirchhoff's junction rule and Kirchhoff's loop rule to obtain the appropriate number of independent equations.
3. Set the current through the bridge equal to zero. Show that this leads to the equations in the introduction to Experiment 19.

Experiment 19: Bridge Circuits

Introduction: A bridge circuit can be used to determine an unknown resistance or capacitance by taking advantage of a symmetric arrangement. Consider the Wheatstone bridge circuit illustrated in Fig. 19-1. The galvanometer deflects when a current travels up or down through the "bridge." A standard resistance and unknown resistance (or standard and unknown capacitances) are placed at the top of the circuit. The bottom resistances are two sections of a long, bare wire. The ratio of the lengths of the two sections can be varied by sliding the tap key. When the tap key is positioned where the galvanometer reads zero, the bridge is said to be *balanced*. This position is easy to find experimentally, and also simplifies the theory.

Consider the balanced position, where the current through the bridge is zero, so $I_s = I_u$ and $I_1 = I_2$. Applying Kirchhoff's loop rule to the left and right loops,

$$I_s R_s = I_1 R_1 \quad , \quad I_s R_u = I_1 R_2$$

Dividing these two equations,

$$\frac{R_u}{R_s} = \frac{R_2}{R_1}$$

The resistances R_1 and R_2 are proportional to the respective lengths of the two sections of the wire separated by the tap key:

$$R_{1,2} = \rho \frac{L_{1,2}}{A}$$

where L_1 and L_2 are the lengths of the two sections and A is the cross-sectional area of the wire. It follows that

$$\frac{R_u}{R_s} = \frac{L_2}{L_1}$$

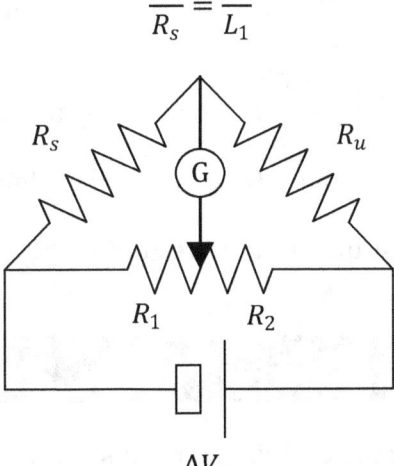

Fig. 19-1. A schematic diagram for the Wheatstone bridge.

Textbook Reading: Read the sections of the textbook about Kirchhoff's rules, resistance, resistivity, and any sections or examples with a similar bridge circuit. The equations for resistivity and Ohm's law, and the strategy for applying Kirchhoff's rules are related to this experiment.

Objectives: Construct a bridge circuit and employ it to determine an unknown resistance (or capacitance).

Apparatus: DC power supply, conducting wire with meterstick and tap key, known resistance (or capacitance), unknown resistance (or capacitance), galvanometer (or voltmeter or oscilloscope), electrical wires.

Procedures: Set up the circuit illustrated schematically in Fig. 19-1. Wait for the instructor to approve your circuit before turning power on.

Caution

Do not exceed the maximum potential difference specified by the instructor. If it smells like the resistor is generating much heat, turn your power supply off and consult the instructor (if you can smell smoke, do not touch the resistor or conducting wire to check this – just turn the power off immediately).

Starting with zero potential difference, turn the DC power supply on and gradually turn up the potential difference to the desired value. Being wary not to touch the bare wire, slide the tap key along the conducting wire. Press it down for a moment to view the deflection of the galvanometer's needle. Locate the position of the tap key necessary to balance the bridge. Find the uncertainty in this position. Record the two lengths of the conducting wire, the uncertainty in the tap key's position, and the known resistance (or capacitance).

Data Check: Prior to leaving lab, the following data should be gathered:
- ✓ known resistance (or capacitance)
- ✓ the uncertainty in the tap key's position
- ✓ the two lengths of the conducting wire, separated by the tap key, including uncertainties

Analysis: Calculate the unknown resistance (or capacitance) along with its uncertainty.

Results: Make a table of the main quantitative results for this experiment. State results in the format $x \pm \sigma_x$.

* * * * * * * * * **What to Turn In** * * * * * * * * * *

The Report: The report for Experiment 19 should include the following sections:
- ✓ Abstract
- ✓ Data Table.
- ✓ Analysis.
- ✓ Tabulation of Results.
- ✓ Conclusions.

Research: Write the algorithm for your research project. Add references and submit an updated References section.

Prelab Exercises for Experiment 20: Complete the following prelab exercises prior to the next lab.

1. For the graph illustrated in Fig. 19-2, read the following directly from the plot. For each, explicitly describe what you read from the plot and how. Do *not* calculate any of the following – just read them directly from the plot.

    a. the baseline – i.e. the minimum value of N
    b. the initial value of N for which there is data
    c. the initial time for which there is data
    d. the half-life; if the baseline and initial time were zero, you would read this as the time it takes to reduce to half of its initial value
    e. the time constant; if the baseline and initial time were zero, you would read this as the time it takes to decay to $1/e$ of its initial value

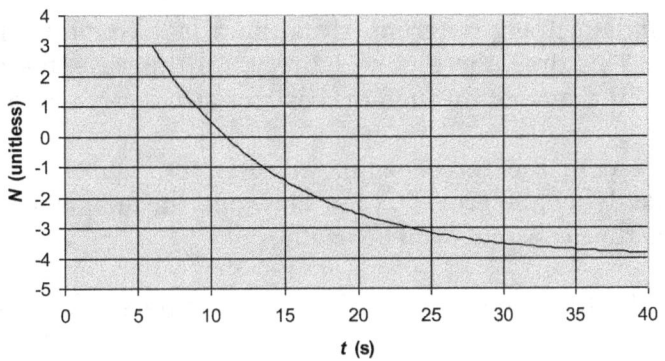

Fig. 19-2. A graph shows exponential decay.

Experiment 20: RC Circuits

Introduction: Consider a resistor and initially uncharged capacitor connected in series with the battery. The charges in the connecting wires are free to flow, except that they cannot pass through the capacitor (except for a dangerously large potential difference that the dielectric between the conductors breaks down, resulting in a spark). The charges can redistribute, though. Electrons will flow (constituting a current), redistributing such that one plate of the capacitor becomes positively charged (the one that is connected to the positive terminal of the battery), and the other becomes negatively charged. The current begins with the maximum value $i_m = \varepsilon/R$, where ε is the emf of the battery, and decays exponentially to zero. The reason for the decrease in current is that the charges are more reluctant to move as the capacitor becomes more fully charged. The charge stored on the capacitor experiences exponential growth, getting closer and closer to its maximum value $q_m = C\varepsilon$.

A similar, yet fundamentally different, situation arises when a fully charged capacitor is connected to a resistor. This time, there is no battery. In this case, the charge on the capacitor exponentially decays, and the current through the resistor also decays exponentially.

Textbook Reading: Read in the textbook about resistors, capacitors, and RC circuits. The equations for exponential decay and growth, Ohm's law and the analogous equation for capacitors, and the equations for the time constants will be useful.

Objective: Investigate the time-dependence of potential difference across a capacitor in an RC circuit.

Apparatus: function generator, RLC circuit board (or resistor, capacitor, and circuit board), voltage sensors (or voltmeters), computer interface, electrical wires.

Procedures: Connect the function generator to the resistor and capacitor in series with one another, but do not turn power on until the instructor approves your circuit. Connect voltage sensors (or voltmeter) to measure the potential difference across each element. Ensure that the polarities are consistent. If using a voltmeter, a combination of resistor and capacitor with a suitable time constant will be needed.

Choose a square wave. This essentially turns a DC power supply on and off repeatedly such that the current will decay for a half-cycle, then grow for a half-cycle, and so on.

The period is the time it takes to complete two half-cycles. Since the period is the reciprocal of the frequency, the frequency will affect how much time the current spends decaying and growing. The decays and growths should have *about* 5 time constants or more to reasonably complete their decays and growths.

Compute the desired frequency as follows. First, determine the time constant from the resistor and capacitor connected in series. Remember to use SI units – i.e. account for any prefixes. Multiply this by 10 to figure out how much time the current should decay or grow (the factor of two is present since a decay by itself is just a half-cycle). This equals one period. Reciprocate this to find the frequency. This is a good starting point, but it may need to be adjusted a little once you have some data.

Set the function generator to the desired frequency. Select the voltage sensor in the data collection software program. The sampling rate needs to be about 100 times greater than the frequency in order to obtain about 50 data points per half-cycle.

Note

For many function generators, the frequency knob is not a multiplier. In this case, the frequency range buttons indicate the maximum frequency for that setting. For example, depressing the 200 Hz button and adjusting the frequency knob to 0.5 gives 50 Hz (*not* 100 Hz).

Wait for your instructor to approve the circuit before turning power on. Beginning with zero amplitude, slowly increase the voltage to the desired value. Check the meter sensitivity to ensure that you stay within its tolerance.

Caution

Do not exceed the maximum amplitude specified by the instructor. If it smells like the resistor is generating much heat, turn your power supply off and consult the instructor (if you can smell smoke, do not touch the resistor to check this – just turn the power off immediately).

Record values for the resistance, capacitance, frequency, and power supply voltage. If using the interface and the time constant is small compared to a second, gather data for just a second or two – if you wait longer, you will have way too much data. View the voltage data with a graph in the data analysis software (later on, after you isolate the data you want to analyze, you will make a graph in Excel). You will need to stretch out the time axis so that you can see within the span of a time constant; otherwise, you will not be able to view the graph properly, and you will be misinterpreting everything. Continue to stretch out the horizontal axis until you see one complete decay and one complete growth on the screen. Decide whether you need to graph the potential difference across the resistor or capacitor in order to obtain both decays and growths.

The decays and growths should look reasonably complete. There should be plenty of data points for each decay and growth. If not, try adjusting the frequency and sampling rate until a good graph is obtained.

Once you obtain a good plot, highlight the data corresponding to one complete decay and save it in Excel. Repeat for a complete growth.

Data Check: Prior to leaving lab, the following data should be gathered:
- ✓ resistance, capacitance, frequency (which may have been changed since the initial setting), and power supply voltage
- ✓ data for a graph of potential difference as a function of time for the decay
- ✓ data for a graph of potential difference as a function of time for the growth

Analysis: Show your calculation of the frequency first used. If this was changed afterward, state the frequency that corresponds to your graphs. Determine the theoretical maximum current in the circuit and the maximum charge stored by the capacitor. Prepare the graphs for exponential decay and growth.

Examine the decay plot. Read the initial and baseline voltages and the initial time directly from the plot. Determine the experimental half-life by reading a time off the graph. You are not looking for the time where the voltage is half the maximum, but the time where the voltage is halfway between the maximum and baseline values of the decay. Then subtract the initial time from this time to obtain the half-life. Compute the theoretical time constant, and use this to calculate the theoretical half-life. Compare the two half-life's.

Determine the experimental time constant by reading a time off the graph. Without vertical or horizontal shifts, this would be the time it takes to decay to $1/e$ of its initial value. Instead, this will be the time elapsed before reaching $1/e$ of its initial value relative to the baseline, after subtracting the initial time. Compare to the theoretical time constant.

Make a similar analysis of the growth plot. In this case, the experimental time constant is based on the time it takes to grow to $(1 - 1/e)$ of its maximum value relative to the baseline, again after subtracting the initial time.

Results: Make a table of the main quantitative results for this experiment. State results in the format $x \pm \sigma_x$. Include relevant percent errors or differences and state the relative discrepancies.

* * * * * * * * * * * **What to Turn In** * * * * * * * * * * *

The Report: The report for Experiment 20 should include the following sections:
- ✓ Data Table.
- ✓ Analysis.
- ✓ Tabulation of Results.
- ✓ Theory.

Research: Begin implementing the algorithm. Include a note that describes what progress you have made on your algorithm thus far. Begin a discussion for your research project. This will be similar to a Discussion of Results, but geared toward your research project. One thing you can do now is describe how your numerical program is being employed. You will need to obtain numerical results in order to complete your discussion, so just write what you can with what you have thus far. If you have no results at this point, you can at least start to make some comparisons and begin other ingredients to your discussion, and then complete the discussion when you obtain results. Since the discussion is one of the heartier sections of the research paper, it is important to make some progress on it now.

Prelab Exercises for Experiment 21: Complete the following prelab exercises prior to the next lab.
1. A 4-m long, straight current-carrying wire lying in the xy-plane carries a 3-A current heading at an angle of 30° counterclockwise from the positive x-axis. There is an 800-G uniform magnetic field directed at angle 45° counterclockwise from the positive y-axis. Find the magnitude and direction of the net force exerted on the current-carrying wire.

Experiment 21: Magnetic Force

Introduction: Mass has two fundamental definitions: In Newton's laws of motion, mass serves as a measure of inertia, while in Newton's law of universal gravitation, mass is the source of a gravitational field. Electric charge is analogous to the second definition of mass in that electric charge is the source of an electric field. Related to the electric field is the magnetic field. What is the source of a magnetic field? Although a magnet might come to mind, it is not a good answer because the magnetic field of a magnet depends on something more fundamental at the microscopic level. Moving charge, or electric current, is the source of a magnetic field. All moving charges, and hence currents, produce magnetic fields (so it is not necessary to have a magnet to produce a magnetic field).

Not only is mass the source of a gravitational field, but when a mass is placed in a gravitational field it experiences a gravitational force (called weight). This force is always along the gravitational field. Similarly, when an electric charge is placed in an electric field, it experiences an electric force. This force is either along the electric field or opposite to it, depending upon the sign of the charge. When a current, or moving charge, is placed in a magnetic field, it experiences a force. However, unlike gravitational and electric forces, magnetic forces are always perpendicular to both the magnetic field and the current – a seemingly strange property of magnetic fields.

Textbook Reading: Read about magnetic fields, magnetic forces, and the right-hand rule for determining the direction of the magnetic force. The equation for the magnetic field produced by a long straight wire and the equation that relates magnetic field to magnetic force will be useful.

Objectives: Experimentally verify that a magnetic force can be exerted on a current-carrying conductor in the presence of an external magnetic field. Determine the magnetic field strength in the central region of a magnet cradle.

Apparatus: current balance, rods and supports, DC power supply, scale, magnet cradle, ammeter, cards with conducting wires and banana plugs, external resistance.

Procedures: You will be typing a set of procedures for today's lab. The manual is providing just a general sense of direction. Your procedures will specify more precisely and thoroughly how to conduct this experiment. It would be wise to write down notes during lab that will be useful for preparing the procedures later – in addition to obvious points that you don't want to overlook later, be sure to record details that do not come to mind easily.

Your goal is determine the magnetic field in the center of a permanent magnet cradle from the slope of a plot. (The plot is part of the analysis, not the procedures, but you need to have the analysis in mind in order to develop the experimental procedures.) You will need one independent and one dependent variable. Let an equation serve as a guide for what you need to measure to be able to compute the magnetic field.

Once you figure out what to measure, you can decide what available equipment can be utilized to make the measurements.

You can use a common, simple device to deduce whether the red or white face of the magnet are magnetic north or south. Be sure to position something in the center of the magnet, or you will be finding the magnetic field somewhere else.

You will need to connect a circuit. As always, do not turn power on until your instructor approves your circuit. Begin with zero current and voltage, and gradually increase them, keeping a careful eye on the meter readings so that you don't blow a fuse. Note the ammeter's maximum, and be sure to stay safely below this threshold.

Caution

Do not exceed the maximum current specified by the instructor. If it smells like the resistor is generating much heat, turn your power supply off and consult the instructor (if you can smell smoke, do not touch the resistor to check this – just turn the power off immediately).

The electronic scale is noticeably sensitive in this experiment. If you lean on the table, for example, you affect the scale readings. So ensure that nobody is touching the table, walking by, etc. while you are making measurements.

(The scale is actually reading the force exerted on the cradle, but by Newton's 3rd law this equals the force that the cradle exerts on the scale.)

If you are using a Pasco current balance for this experiment, examine the cards with the conducting wires carefully: The conducting wire continues onto the back of a couple of these cards, which must be accounted for; otherwise, the results can be off by a significant factor.

Data Check: Prior to leaving lab, the following data should be gathered:
- ✓ enough data for a good linearized plot, from which the magnetic field strength can be determined in terms of the slope

Analysis: Compute the magnitude of the magnetic field in the central region of the magnet in terms of the slope of a linearized plot of the data. Also, calculate its uncertainty.

Results: Make a table of the main quantitative results for this experiment. State results in the format $x \pm \sigma_x$.

* * * * * * * * * * **What to Turn In** * * * * * * * * * * *

The Report: The report for Experiment 21 should include the following sections:
- ✓ Abstract.
- ✓ Data Table.
- ✓ Analysis.
- ✓ Tabulation of Results.
- ✓ Experimental.

Research: Continue working on your algorithm. Make some preliminary graphs of your numerical data, based on what your program is able to do thus far. As always, describe each graph in the caption.

Prelab Exercises for Experiment 22: Complete the following prelab exercises prior to the next lab.

1. A 5-cm long solenoid with 1-cm diameter and 80 turns carries a current of 300 mA. What is the magnetic field in the center of the solenoid?

Experiment 22: Magnetic Field

Introduction: Current-carrying conductors produce magnetic fields. When a DC power supply is connected to a solenoid – which has the structure of a wire wrapped around a right-circular cylinder – the current running through the numerous loops reinforce one another inside the solenoid. The magnetic field in the center of the solenoid is proportional to the number of turns (loops), and also depends upon the current and the length of the solenoid.

Textbook Reading: Read the sections in the textbook on magnetic fields, Ampère's law, and solenoids. The equation for the magnetic field near the center of a solenoid is needed in the analysis.

Objectives: Estimate the number of loops in a finite, tightly wound solenoid from geometry. Compare this to a result obtained from measurement of the magnetic field in the central region of the solenoid connected as part of an RL circuit.

Apparatus: DC power supply, RLC circuit board (or a solenoid and resistor), ammeter, electrical wires, magnetic field sensor, vernier calipers, ruler.

Procedures: Estimate the number of loops of wire in the finite, layered solenoid by simple geometry (without connecting any circuit). Record the necessary data, including relevant length measurements. Remember to include uncertainties. Enter the scratch work for your calculation in the notebook. Bear in mind that you will be asked to describe this calculation in detail later as well as draw and label a diagram.

Connect the DC power supply to the resistor and inductor in series with one another and the ammeter. Wait for your instructor to approve the circuit before turning power on.

Caution

> Do not exceed the maximum current specified by the instructor. If it smells like the resistor is generating much heat, turn your power supply off and consult the instructor (if you can smell smoke, do not touch the resistor to check this – just turn the power off immediately).

Record data that will enable you to compute the number of loops of wire in the solenoid, using your electricity and magnetism data, from the slope of a graph.

It would be wise to check if the two results agree reasonably before leaving lab. If not, check over both parts of the experiment carefully, redoing either experiment, if necessary.

Data Check: Prior to leaving lab, the following data should be gathered:
- ✓ length measurements with uncertainties for geometry method
- ✓ electricity and magnetism data for magnetic field method

Analysis: Use your geometric data to compute the number of loops in the solenoid. Also, calculate the uncertainty in the number of loops.

Make a linearized plot using your electricity and magnetism data. Determine the number of loops in the solenoid in terms of the slope of this plot, and also establish its uncertainty.

Results: Make a table of the main quantitative results for this experiment. State results in the format $x \pm \sigma_x$. Include the percent error or difference and state the relative discrepancy.

* * * * * * * * * *   What to Turn In   * * * * * * * * * *

The Report: The report for Experiment 22 should include the following sections:
- ✓ Data Table.
- ✓ Analysis.
- ✓ Tabulation of Results.
- ✓ Illustration.
- ✓ Theory.

Research: Complete your numerical algorithm. Type an abstract for your research paper, based on what you have thus far. Complete your Discussion section, but save conclusions for a separate Conclusions section (assigned in the next lab).

Prelab Exercises for Experiment 23: Complete the following prelab exercises prior to the next lab.
1. Design an experiment to test Faraday's and Lenz's laws. Include:
    a. a list of equipment that is likely to be available; some common household materials, such as aluminum foil, can be brought to the lab, if needed
    b. a set of procedures that thoroughly describe the details of how the experiment will be conducted, including which quantities will be measured and how the measurements will be made; this may include qualitative observations
    c. a description of how the needed analysis will be performed

Experiment 23: Faraday's Law

Introduction: According to Faraday's law, a current can be induced in a loop of wire when the net magnetic flux through the area of the loop is changing. Magnetic flux is a measure of the relative number of magnetic field lines passing through a surface – in this case, the area of the loop. Three ways to change the magnetic flux through the area of the loop include: increasing or decreasing the magnetic field in the area of the loop, changing the area of the loop, or rotating the loop relative to the magnet. Lenz's law determines the direction of the induced current.

Textbook Reading: Read about magnetic flux, Faraday's law, and Lenz's law in the textbook. The equations for magnetic flux and Faraday's law, as well as the strategy for applying Lenz's law, will be useful during this experiment.

Objective: Design and conduct an experiment to test Faraday's and Lenz's laws.

Apparatus: equipment and supplies for investigating Faraday's and Lenz's laws.

Procedures: Once you obtain the approval of the instructor, carry out the experiment you designed as a prelab exercise. If your experiment involves an electric circuit, leave the power off and seek the approval of your instructor before turning power on.

Data Check: Prior to leaving lab, the following data should be gathered:
- ✓ Make a list of what data you need in order to meet the objectives of your experiment. Record qualitative observations as well as quantitative data. For qualitative data, include uncertainties.

Analysis: Perform any calculations with quantitative data that are needed in your analysis of Faraday's and Lenz's laws.

Results: Make a table of the main quantitative (or qualitative) results for this experiment. State any quantitative results in the format $x \pm \sigma_x$.

* * * * * * * * * * **What to Turn In** * * * * * * * * * * * *

The Report: The report for Experiment 23 should include the following sections:
- ✓ Data Table.
- ✓ Analysis.
- ✓ Tabulation of Results.
- ✓ Experimental.
- ✓ Discussion of Results.

Research: Check your algorithm for mistakes and check your numerical results for consistency, that they make sense, and that they agree with theoretical examples or behave appropriately in certain limits. Type a Conclusions section for your research paper.

Prelab Exercises for Experiment 24: Complete the following prelab exercises prior to the next lab.

1. Find all of the formulas alluded to in the introduction to Experiment 24. For each equation in your list, briefly describe its significance/usefulness, define each symbol, and indicate the SI units for each symbol. Include a list of reliable references used.

Experiment 24: RLC Circuits

Introduction: Consider a resistor, inductor, and capacitor wired in series in an AC circuit. The potential differences across resistors, inductors, and capacitors do not combine in the usual way in an AC circuit. This originates from the fact that the potential difference of the power supply is, by definition, time-dependent; and, for a sinusoidally varying emf produced by the power supply, the time-dependent current that it produces lags behind the potential difference across the inductor by 90° (as a result of Faraday's law) and leads the potential difference across the capacitor by 90°. The potential difference across the resistor, however, is in phase with the current. If the current in the circuit behaves like a sine wave, the potential differences across the inductor and capacitor behave like negative and positive cosine waves, respectively; the difference is that the cosine curve differs from the sine curve by a horizontal shift of 90°. Recall that when sine is maximum (equal to 1, which occurs at 90°), cosine is zero, and when cosine is maximum (it is 1 at 0°), sine is zero. Similarly, when the potential difference across the resistor is zero, the potential difference across the inductor is its most negative value and the potential difference across the capacitor is maximum; when the potential difference across the resistor is maximum, the potential differences across the inductor and capacitor are zero.

We say that the potential differences are out of phase with the current by 90° for the inductor and capacitor. Since they are not in phase, when applying Kirchhoff's loop rule to combine potential differences across circuit elements, the root-mean-square (rms) voltages that an AC voltmeter measures across the circuit elements cannot simply be added together in the usual way. However, taking into account the phase angle, the potential differences can be added like vectors. In this context, the potential differences are promoted to phasors – quantities that behave like vectors, with a magnitude and direction. Phase addition is identical to vector addition.

Ohm's law still applies to a resistance even in an AC circuit; it applies both to relate the rms current and voltage as well as the instantaneous values. A similar equation can be written for the inductor, but only for the rms values. In this case, the proportionality factor between the root-mean square voltage and current is called the inductive reactance. There is an analogous capacitive reactance pertaining to the capacitor. In addition, the root-mean square voltage across the power supply is proportional to the root-mean square current by a factor called the impedance, which can be related to the resistance, inductive reactance, and capacitive reactance via phasor addition.

It turns out that the reactances depend on the frequency of the AC power supply, which means that the impedance depends on the frequency. Thus, for a given rms voltage, the rms current depends on the frequency. The frequency that maximizes the rms current is called the resonance frequency.

Textbook Reading: Read the chapter in the textbook on AC circuits, including resistors, inductors, and capacitors in AC circuits, phasors, phasor addition, RLC circuits, and resonance. The equations for inductive reactance, capacitive reactance, impedance, Ohm's law, and resonance frequency will be useful in the analysis.

Objectives: Observe resonance in an RLC circuit. Compute the resonance frequency two different ways.

Apparatus: Function generator, RLC circuit board (or separate resistor, inductor, and capacitor), AC voltmeter (or oscilloscope), electrical wires.

Procedures: Connect the function generator (or other sinusoidal AC power supply) in series with the resistor, inductor, and capacitor. Record the resistance, inductance, and capacitance.

Set the waveform to sine wave and set the frequency. Measure the rms voltage across the power supply, resistor, inductor, and capacitor. Record these values and the frequency. Choose a logarithmic range of frequencies (from about 1 Hz to 20 kHz) and repeat for each frequency.

Data Check: Prior to leaving lab, the following data should be gathered:
   ✓ resistance, inductance, and capacitance values
   ✓ rms voltage across the power supply, resistor, inductor, and capacitor for each of several frequencies

Analysis: Make a plot of $\Delta V_L - \Delta V_C$ as a function of frequency. Choose a logarithmic scale for each axis to make this a log-log plot. The data points are expected to form two straight lines, which intersect at the resonance frequency. Draw these lines and find the point of intersection in order to determine the resonance frequency. Split the data points up into two groups and plot two different series on one graph in order to fit trendlines to each set. Also, compute the resonance frequency from a formula and compare.

For each frequency, compute the experimental ratios $\Delta V_C/\Delta V_R$ and $\Delta V_L/\Delta V_R$. Check for agreement by substituting in the formulas for the reactances and resistance and cancelling the currents. Verify that ΔV_R, ΔV_L, and ΔV_C do not add up to the power supply voltage, but that the vector addition of the corresponding phasors does have a magnitude equal to the power supply voltage.

Results: Make a table of the main quantitative results for this experiment. State results in the format $x \pm \sigma_x$. Include percent errors or differences and state the relative discrepancies.

* * * * * * * * * * **What to Turn In** * * * * * * * * * * *

The Report: The report for Experiment 24 should include the following sections:
   ✓ Data Table.
   ✓ Analysis.
   ✓ Tabulation of Results.
   ✓ Theory.

Research: Submit your completed paper. Include any necessary updates and revise your paper based on any feedback that you have already received. Exchange your paper with a peer. Review your peer's paper thoughtfully and carefully. Make some notes on your peer's draft that *suggest* conceptual or structural changes that *might* be useful. You are not grading the paper, nor are you stating what is correct or incorrect. Be suggestive. Express your constructive comments collegially. Be sure to include a few positive comments for which aspects of the paper you like the most.

Prelab Exercises for Experiment 25: Complete the following prelab exercises prior to the next lab.

1. What is the conceptual significance of index of refraction?
   a. How is it defined mathematically?
   b. State an inequality that indicates what values index of refraction may have, physically.
   c. Explain why the index of refraction cannot have any value.
   d. Predict the index of refraction for a common glass slab. Include a reliable reference to the source of your information.
   e. Using your answer to (d), what is the speed of light in such glass?
   f. Use your answer to (d) again for this problem. A ray of light in air is incident upon a glass slab. The incident ray makes an angle of 35° with the normal. Predict the angle of the refracted ray.

Experiment 25: Index of Refraction

Introduction: When a ray of light in air is incident upon a glass slab, the ray refracts through the glass, bending toward the normal where the speed of light is effectively less. When it exits the glass, it again refracts, this time bending away from the normal as the speed of light effectively increases. In either case, the refraction is described by Snell's law.

Textbook Reading: Read the sections in the textbook on refraction, index of refraction, and Snell's law. The equation for Snell's law and the equation that defines index of refraction will be useful.

Objective: Determine the index of refraction of a block of glass.

Apparatus: block of glass, corkboard, dissecting pins, blank white paper, ruler, protractor. A refraction tank also works well.

Procedures: Fasten a blank white sheet of paper to a corkboard. Place a block of glass on the center of the paper. Carefully trace the outline of the glass. Be careful not to adjust the position of the glass until the experiment is complete. If you pick the glass up, you will need to find a new page and draw a new rectangle.

Place two pins on one side of the block such that the line passing through the pins represents an incident ray of light. Choose the position of these pins carefully. Imagine this incident beam refracting through the glass and emerging through the opposite face – this will only occur with proper pin placement. Angles that are too close to perpendicular or parallel to the glass slab will lead to more experimental error – so avoid these extreme angles. Also, you want a variety of angles – you want to avoid accidentally repeating nearly the same angle. To achieve the best variety of angles, place one pin near the edge of the glass – as opposed to the middle of one side – and leave it fixed (call it the "hingepin"). There is no reason to remove the glass or the hingepin until the entire experiment is complete.

Look through the opposite side of the glass. With proper pin placement, you will be able to position your eye somewhere such that the two pins line up – it will look like one pin blocks the other pin. Now place two pins on this opposite side of the block so that it looks like all pins line up when you look through the glass. Line up the bottoms of the pins, not the tops. Why? Although the pins will appear to line up when you look through the glass – the first pin will appear to block the other three out of view – the pins will not appear to line up from a top view because of refraction through the glass.

Gather five sets of four pins that will form a good plot. Label each set as you gather it. Ensure that your incident angles are sufficiently different from one another.

Now remove the glass. For each set of four pins, draw two incident rays (one on each side of the glass) and the connecting refracted ray. Obviously, the refraction must be drawn at the outline of the glass, where it occurs. Measure the two incident and two refracted angles for each set. Measure all angles to the nearest tenth of a degree by estimating the last digit. Remember that the angles are defined relative to the normal. There is less error, however, in measuring them relative to the surface and then subtracting to obtain the needed angles.

Data Check: Prior to leaving lab, the following data should be gathered:
- ✓ ray diagram showing the refraction of five rays through the glass, including the outline of the glass
- ✓ five sets of incident angles on either side of the glass
- ✓ five sets of angles of refraction on either side of the glass

Analysis: Use the data to make a linearized plot. Determine the index of refraction of the glass in terms of the slope. Also, find the uncertainty in the index of refraction.

Results: Make a table of the main quantitative results for this experiment. State results in the format $x \pm \sigma_x$.

* * * * * * * * * * **What to Turn In** * * * * * * * * * * * *

The Report: The report for Experiment 25 should include the following sections:
- ✓ Abstract.
- ✓ Data Table.
- ✓ Analysis.
- ✓ Tabulation of Results.
- ✓ Conclusions.

Research: Get together with the peer whose paper you reviewed and exchange your ideas in a friendly, collegial manner. Be positive and suggestive. Point out what is good as well as what you feel could use improvement. Your peer will make similar comments on your paper. Do not take this as a personal attack. You may not agree with the points, so remember that these are suggestions. Appreciate the time your peer took to review your paper, and later choose which advice, if any, to take. Revise your paper.

Prelab Exercises for Experiment 26: Complete the following prelab exercises prior to the next lab.
1. A 6-cm tall object is place 24 cm before a convex lens with 40-cm focal length.
   a. Where does the image form?
   b. How tall is the image?
   c. What is the orientation of the image?
   d. What is the character of the image?
   e. What is the magnification of the image?

Experiment 26: Thin Lenses

Introduction: A lens is thin if its thickness is small compared to its diameter. A thick lens requires some care to account precisely for the refraction through the lens, whereas a thin lens allows for convenient approximations with very little error. A lens that causes incident parallel rays of light to converge is called a converging lens, while a lens that causes incident parallel rays of light to diverge is called a diverging lens. Of course, there are many situations in which rays of light are incident upon a lens, but in which the rays are not parallel – such is the case in today's lab. A double-convex lens or plano-convex lens, for example, is converging, while a double-concave lens or plano-concave lens is diverging. A lens that has one concave and one convex side could be either, depending upon the curvature of the two sides. Sometimes a double-convex lens is simply called convex, and similarly with double-concave.

When an object is placed before a lens, an image forms. Where the image forms, as well as the size, orientation, and character of the image, depend upon the position of the object, the focal length of the lens, and the nature of the lens (i.e. convex or concave). A real image can only be formed for a single convex lens if the distance between the screen and image is greater than or equal to four focal lengths.

π π π π π π π π π π π π π π π π π π π π π π π π π π π π π π π π π π π π π π π π π π

π **Proof**: The focal length f equals $f = pq/(p + q)$, where p and q are the object and image π

π distances, respectively. The numerator can be rewritten as: $pq = [(p + q)^2 - (p - q)^2]/4$. π

π Thus, $f = [(p + q)^2 - (p - q)^2]/4(p + q)$ or $f = (p + q)/4 - (p - q)^2/4(p + q)$. This shows π

π that the focal length is smaller than $(p + q)/4$, since a positive number is subtracted from it, π

π for the case of a real image (requiring $q > 0$). Since $p + q$ equals the distance between the π

π screen and the object, a real image can only be produced if $(p + q) \geq 4f$. π

π π π π π π π π π π π π π π π π π π π π π π π π π π π π π π π π π π π π π π π π π π

The equations for thin lenses follow some sign conventions. Following is a popular sign convention:

- For a single lens system, the object distance and object height are always positive.
- A positive image distance corresponds to an image that forms on the opposite side of the lens compared to the object.
- The image height is positive for an upright image and negative for an inverted image. Whether it is upright or inverted is relative to the object.

The orientation of the image is whether or not it is upright. The character of the image is whether the image is real – rays of light actually pass through the image – or virtual – rays of light diverge from a point that can be traced back to the image, but these rays do not actually pass through the image. A singe concave lens, for example, forms a virtual image. The character is real if the image distance is positive.

Textbook Reading: Read the sections of the textbook on thin lenses, ray tracing, and image formation. The equation for focal length and two equations for magnification will be useful. The strategy for solving problems with two lenses will also be needed in the analysis.

Objectives: Determine the focal lengths of various lenses. Learn how the conjugate positions are related for a fixed object and screen.

Apparatus: light source, optic bench, optic mounts, lens holders, lenses, object, screen, ruler, meter stick, flashlight.

Procedures: Begin with a converging lens. Use a light source, object, and screen to locate an image. The image has been correctly found when it is sharply in focus (otherwise, it will appear blurry). Record a range of values over which the image appears equally sharp in order to establish the best value and its uncertainty. Hold up a meterstick or ruler to measure distances directly since one or more optical elements may be off-center compared to its optical mount. Record the object distance, image distance, object height, image height, orientation, and character of the image.

Leave the screen and object fixed and move the lens to find the conjugate position. That is, for a fixed screen and object, there are actually *two* positions where a lens can be placed and obtain an image; find this second position, and record information for this conjugate position.

Perform a quick scratch calculation to ensure that the results are plausible. For example, if the distance between the screen and object does not exceed four times the focal length, then the experiment must be repeated.

Add a diverging lens. Use the same converging lens together with the diverging lens so that the focal length of the converging lens, which you can now calculate, can be used to determine the focal length of the diverging lens. This is necessary since a single diverging lens by itself cannot produce a real image. Locate a real image for this lens system. It may be necessary to adjust the distance between the lenses, the position of the object, the position of the screen, or the order of the lenses in order to do this (since some combinations will lead to virtual images instead). A little math can offer some guidance. Once a focused image is located, record the object distance for the first lens, the image distance for the second lens, the object and image heights, and the orientation and character of the final image.

Data Check: Prior to leaving lab, the following data should be gathered:
- ✓ object and image distances and heights for a single converging lens
- ✓ orientation and character of the image for the single converging lens
- ✓ object and image distances and heights for the conjugate position
- ✓ orientation and character of the image for the conjugate position
- ✓ object and image distances and heights for the lens system
- ✓ orientation and character of the image for the lens system

Analysis: Compute the focal length, and compute the magnification using two different equations. Do this for each of the two conjugate positions for each lens.

Results: Make a table of the main quantitative results for this experiment. State results in the format $x \pm \sigma_x$. Include relevant percent errors or differences and state the relative discrepancies. Also, include qualitative features in the table – namely, the character and orientation of the image.

* * * * * * * * *   What to Turn In   * * * * * * * * * *

The Report: The report for Experiment 26 should include the following sections:
- ✓ Data Table.
- ✓ Analysis.
- ✓ Tabulation of Results.
- ✓ Introduction.
- ✓ Sources of Error.

Prelab Exercises for Experiment 27: Complete the following prelab exercises prior to the next lab.
1. Describe the problem-solving strategy in detail for a problem where an object is placed before a system consisting of two lenses.
2. For a typical telescope, how does the initial object distance compared to the final image distance?

Experiment 27: Telescopes

Introduction: Two lenses can form an effective telescope, depending upon how they are utilized. Such a telescope is useful for viewing objects that are very far away in that it magnifies the apparent size of a viewed object. The viewer looks through the eyepiece; the other lens is called the *objective*. Depending upon the combination of lenses used, the telescope may produce an upright or inverted image.

Textbook Reading: Read the sections in the textbook on lens systems and telescopes. It will be useful to know how apparent magnification relates to the apparent size of an observed object.

Objectives: Use different combinations of lenses to build telescopes. Construct a variety of different telescopes and make comparisons in order to determine:
- a rule for predicting how far the two lenses should be separated
- a rule for determining which lens should serve as the eyepiece and which as the objective
- a rule for predicting the magnification of the telescope
- how these rules are affected if one of the lenses is concave
- the difficulties in obtaining higher and higher magnifications

Apparatus: short optic bench, optic mounts, lens holders, lenses, ruler, meter stick.

Procedures: Through trial and error, build systems of two lenses. For each pair of lenses, vary the distance between them until a focused image can be obtained when viewing a distance object. Determine which lens should serve as the eyepiece in order for the telescope to be useful. Estimate the apparent magnification. Continue examining telescopes until the objectives have been satisfied.

Data Check: Prior to leaving lab, the following data should be gathered:
- ✓ qualitative observations for each telescope
- ✓ the distance between the lenses for each telescope, with uncertainties
- ✓ the apparent magnification for each telescope

Analysis: Develop a formula that predicts the distance between two lenses of given focal lengths needed to construct a useful telescope, as well as a formula for what the apparent magnification of the telescope will be, based on patterns observed in your data.

Results: Make a table of the main quantitative results for this experiment. State results in the format $x \pm \sigma_x$.

* * * * * * * * * * **What to Turn In** * * * * * * * * * *

The Report: The report for Experiment 27 should include the following sections:
- ✓ Data Table.
- ✓ Analysis.

- ✓ Tabulation of Results.
- ✓ Illustration.
- ✓ Discussion of Results.

Prelab Exercises for Experiment 28: Complete the following prelab exercises prior to the next lab.

1. Look up all of the equations that will be needed in the analysis for Experiment 28. For the list of equations, briefly describe the significance/usefulness of each, define each symbol, and label each distance in a diagram. Include a reliable reference to the source of your information.

Experiment 28: Diffraction and Interference

Introduction: The wavelike aspect of light is revealed when a coherent, monochromatic light source passes through a narrow opening. Passing through a single slit, light forms a diffraction pattern on a screen, consisting of a series of bright bands separated by short dark spots; when passing through a pair of closely-spaced slits, an interference pattern results (named such since there is interference from light passing through each slit), consisting of a series of bright and dark spots called *fringes*.

Textbook Reading: Read the sections in the textbook on the conditions for constructive and destructive interference, single-slit diffraction, and double-slit interference. The conditions for constructive and destructive interference are related to the patterns observed, and the equations for the positions of the bright and dark spots produced for single-slit diffraction and double-slit interference will be needed in the analysis.

Objectives: Observe the single slit-diffraction and multiple-slit interference patterns. Compute the wavelength of laser light. Thoroughly determine how the features of the patterns depend upon the number of slits; for example, what the single slit and double slit patterns have in common, and predicting the number of weak fringes that appear between bright fringes.

Apparatus: laser, optic bench, single and multiple slits, slit holders, screen, meterstick, ruler, blank paper, tape, flashlight.

Procedures: Avoid turning the laser on and off frequently. Most lasers have a switch to block the laser light while the unit remains on so that this may be avoided. Record the type of laser used (e.g. He-Ne).

Caution

> Do not look into the laser. Beware of stray laser light from other lab stations, and be careful that your direct laser light does not shine toward another lab station or strike a highly reflective surface. However, viewing laser light that reflects from a screen or scatters off-line from air molecules is okay.

For each pattern that is produced on the screen, place a sheet of paper over the screen and trace all the fringes. Moving the slit slide slightly may help to produce a pattern with better contrast. For multiple slits, insure that laser light is passing through all of the slits. Record distances that will be needed to determine the wavelength, including their uncertainties.

Produce single-slit diffraction and double-slit interference for equal slit widths to determine the effect of the second slit. Note which features are the same and which are different for the two cases. See what the effect is of adding a third and fourth slit of the same width. Again, record which features are the same and which are different. Look for patterns in features that differ.

Now examine the effect of changing the slit width on the diffraction pattern. Finally, see what happens to the double-slit interference pattern when the slit spacing changes.

When making a distance measurement from the sketched fringes, using $m = 1$ results in a large relative error since the distance is quite small, while it is easy to make a counting mistake when m is very large – especially, when there are missing orders involved. Missing orders do need to be counted when the distance passes over one. For multi-slit patterns with weak in-between fringes, count only the bright fringes.

Data Check: Prior to leaving lab, the following data should be gathered:
- ✓ sketches of the various single- and multi-slit patterns
- ✓ distance measurements that will permit the wavelength to be calculated for each pattern observed

Analysis: Compute the wavelength of the laser light, including its uncertainty, for the various slit combinations used. Look up the wavelength of the laser light for the laser used from a reliable source of information. You will need to cite the reference. Find the percent error and state the discrepancy in terms of the uncertainty.

Results: Make a table of the main quantitative results for this experiment. State results in the format $x \pm \sigma_x$. Include percent errors or differences and state the relative discrepancies.

* * * * * * * * *   **What to Turn In**   * * * * * * * * * *

The Report: The report for Experiment 28 should include the following sections:
- ✓ Abstract.
- ✓ Data Table.
- ✓ Analysis.
- ✓ Tabulation of Results.
- ✓ Conclusions.

Appendix A: Linear Regression

Introduction: In several experiments the data is plotted and the slope is utilized to compute results. It is better to compute results from the slope – the graphical average – than to compute them for each data point separately and then average the results. The graph also offers greater insight to the behavior of the data.

There are several good objections to making the graphs by hand: The results depend on how accurately the graphs are drawn, and it is necessary to interpolate to compute the slope. Also, there is the problem of how to draw a best fit line. Then these problems are compounded when trying to draw the maximum and minimum slopes to find the uncertainty. Isn't there a better way?

Yes. It is called linear regression. The method of least squares provides formulas for how to draw the best fit and how to compute the slope and its uncertainty. Excel uses such a technique to draw the best fit and to provide the slope, but it does not draw the maximum and minimum slopes nor give the uncertainty. However, Excel does provide a coefficient of linear correlation. Is this related to the uncertainty?

Method of Least-Squares: Linear regression is a useful technique for analyzing data that are expected to obey a linear relationship,

$$y = mx + b$$

where x and y are physical quantities. A plot of y vs. x should result in a straight line with slope m and y-intercept b. The problem is that y and x depend on measurements, and therefore have inherent uncertainties. Thus, the graph will not be perfectly linear. The questions are (1) how to determine if the relationship is indeed linear and (2) what values to use for m and b and their uncertainties. This problem can be approached by applying the principle of maximum likelihood.

If the relationship were perfectly linear, then $y - mx - b$ would be zero for each pair of points (x_i, y_i). The method of least squares sums $(y_i - mx_i - b)^2$ and determines which values of m and b minimize this sum. Here are least-squares method's predictions for m and b:

$$m = \frac{N(\Sigma x_i y_i) - (\Sigma x_i)(\Sigma y_i)}{\Delta}$$

$$b = \frac{(\Sigma x_i^2)(\Sigma y_i) - (\Sigma x_i)(\Sigma x_i y_i)}{\Delta}$$

where

$$\Delta = N(\Sigma x_i^2) - (\Sigma x_i)^2$$

The Greek symbol Σ (capital sigma) represents a sum, for example,

$$\Sigma x_i = x_1 + x_2 + \ldots + x_N$$

and N is the number of data points. The best fit line is then $y = mx + b$. The uncertainties in m and b are given by:

$$\sigma_b^2 = \frac{\sigma_y^2(\Sigma x_i^2)}{\Delta}$$

$$\sigma_m^2 = \frac{N\sigma_y^2}{\Delta}$$

where

$$\sigma_y^2 = \frac{1}{N-2}\left[\Sigma(y_i - b - mx_i)^2\right]$$

The coefficient of linear correlation is

$$r = \frac{\Sigma[(x_i - \bar{x})(y_i - \bar{y})]}{[\Sigma(x_i - \bar{x})^2 \Sigma(y_i - \bar{y})^2]^{0.5}}$$

This coefficient is useful for establishing whether or not x and y obey a linear relationship, or if they are even correlated. If the relationship is perfectly linear, $r = \pm 1$, and if x and y are uncorrelated, r tends toward zero.

L (m)	T (s)
0.20	0.89
0.40	1.23
0.60	1.58
0.80	1.76
1.00	1.99

Table A-1. Data points for which L is proportional to T^2.

Example A-1: Consider the data tabulated in Table A-1, for which L is proportional to T^2. We first linearize the data (as described in the note on page 61), choosing L to be the y-axis variable and T^2 as the x-axis variable (and not T, since L is linear not proportional to T in this example). Thus, the L column is the y column, but we need to make a column where the T data points are squared to serve as the x column. After identifying x and y, we can apply the equations above to determine the uncertainty in the slope – the main result from the linear regression that is often not readily available from spreadsheet applications.

Perform a linear regression on the columns of x and y. This can be achieved by writing a program to carry out the calculations above or using a spreadsheet program like Microsoft Excel. Check that the program works by comparing with the results of Table A-2 (where additional significant figures are shown for the purpose of helping you check the accuracy of your program).

Quantity	Numerical Result
Δ	31.49084494 s^4
m	0.251523896 m/s^2
b	0.003430594 m
r^2	0.996122661
σ_y^2	0.000516979 m^2
σ_m	0.009060019 m/s^2
σ_b	0.023773125 m

Table A-2. The results of performing a linear regression on the data in Table A-1.

Notes

Notes

Notes